T0221265

Routledge Re

Policies and Plans for Rural People

This edited collection, first published in 1988, was the first title to bring international perspectives into the field of rural planning. Using a comparative approach and a broad range of case studies, including Britain, Scandinavia, the U.S.S.R. and New Zealand, the authors review the major problems faced within rural areas, and policy responses to these problems. Each study deals with the political and institutional frameworks involved in the management of rural areas and the means by which policies have been implemented. With an introduction from Paul Cloke that places rural policies and plans within the context of the state, this reissue will be of great value to any students with an interest in the planning and organisation of rural communities across the world.

Policies and Plans for Rural People

An International Perspective

Edited by

Paul Cloke

Routledge
Taylor & Francis Group

First published in 1988
by Unwin Hyman Ltd

This edition first published in 2013 by Routledge
2 Park Square, Milton Park, Abingdon, Oxon, OX14 4RN

Simultaneously published in the USA and Canada
by Routledge
711 Third Avenue, New York, NY 10017

Routledge is an imprint of the Taylor & Francis Group, an informa business

Publisher's Note
The publisher has gone to great lengths to ensure the quality of this reprint but
points out that some imperfections in the original copies may be apparent.

Disclaimer
The publisher has made every effort to trace copyright holders and welcomes
correspondence from those they have been unable to contact.

A Library of Congress record exists under LC control number: 87013070

ISBN 13: 978-0-415-71457-0 (hbk)
ISBN 13: 978-1-315-88257-4 (ebk)
ISBN 13: 978-0-415-71458-7 (pbk)

POLICIES AND PLANS FOR RURAL PEOPLE

An international perspective

Edited by

Paul Cloke

London
UNWIN HYMAN
Boston   Sydney   Wellington

Published by the Academic Division of
Unwin Hyman Ltd
15/17 Broadwick Street, London W1V 1FP, UK

Allen & Unwin Inc.,
8 Winchester Place, Winchester, Mass. 01890, USA

Allen & Unwin (Australia) Ltd,
8 Napier Street, North Sydney, NSW 2060, Australia

Allen & Unwin (New Zealand) Ltd in association with the Port Nicholson Press Ltd,
60 Cambridge Terrace, Wellington, New Zealand

First published in 1988

British Library Cataloguing in Publication Data

Policies and plans for rural people: an
international perspective.
1. Rural development — Government policy
2. Regional planning
I. Cloke, Paul J.
307.7'2   HN49.C6
ISBN 0–04–711017–1

Library of Congress Cataloging-in-Publication Data

Policies and plans for rural people.
Bibliography: p.
Includes index.
1. Rural development—Cross-cultural studies.
I. Cloke, Paul J.
HN49.C6.P65   1988   307.1'4   87–13070
ISBN 0–0–04–711017–1   (alk. paper)

Set in 11 on 11½ point Bembo by BookEns, Saffron Walden, Essex
and printed in Great Britain by Billing and Sons, London and Worcester

Preface

Over recent years there has been a growing awareness of the changes occurring in rural areas, the differential impacts of these changes between different people and different areas, and the role of policies and plans in the state sector which attempt to respond to these impacts. Encouraging strides have been made in our research-based understanding of the context and content of policy-making for rural areas within particular nation states, but at the same time there is a danger of parochialism unless international cross-fertilization of knowledge and insight in these areas can be achieved. It is hoped that this book will make a small initial contribution to that aim. It is the first of two collections of international essays on the subject of rural policies and plans. The focus in this case is on rural people, and a late partner-volume deals with rural land, although this division between people and land is borne of pragmatic convenience rather than any suggestion of mutual exclusivity between the subject matters concerned.

Acknowledgements

Many people have contributed to the task of putting this book together. Each individual author will doubtless acknowledge various sources of help, but for my part I would like to record my sincere thanks to the following: Roger Jones for his enthusiastic support for this and other projects; Robin Hambleton for granting permission to use Figure 1.1; Rob Flynn for his helpful comments on one of the chapters; Maureen Hunwicks, whose accuracy, speed and cheerful disposition render a personal word processor redundant; Trevor Harris for cartographic skills; and lastly my family, Viv, Elizabeth and William, for their constant support and love.

P.J.C.
Lampeter

Contents

List of Tables

1 Introduction: Planning, policy making and state intervention in rural areas

PAUL CLOKE

Introduction

> Planning . . . is an extremely ambiguous and difficult word to define. Planners of all kinds think that they know what it means; it refers to the work they do. The difficulty is that they do all sorts of different things, and so they mean different things by the word; planning seems to be all things to all men. (Hall 1974: 3)

Other than its failure to recognize the broader gender relations within planning and amongst planners, Peter Hall's description of planning as 'all things to all men' provides a very accurate commentary on the contemporary analysis of planning and policy making in rural areas. The chequered present stems directly from a chequered past. Historically, most nations have viewed planning problems as specifically urban phenomena, and whether for reasons of reparation of war damage, or of the sheer political need to cope with the demands of expanding yet decaying cities, planning has been developed as a response to conflict in these urban environments.

The ability of rural interests to attract attention in the political planning arena has therefore depended on the availability of high-profile political issues from which to develop wider mechanisms of planning and policy making. In densely populated nations, planning focus on urban expansion has inevitably led to consideration of rural land-use issues, even if 'rural' in this case could roughly be translated as green spaces on maps into which urban centres could expand. Equally important in the portfolio of rural interests is the question of agricultural production. If governments seek to subsidize agriculture so as to ensure security of production, they are likely to underwrite such policies by giving the protection of agricultural land central importance in rural planning strategies. Thus the preservation of the agricultural land base has in many nations become the fundamental premise of planning for rural areas.

In other, less densely populated nations, this convergence of policies for urban expansion and policies for rural land budgets is less direct and occurs over a longer term. The work of Bill Lassey (1977) and others in the USA has shown that the size and scale of rural territory have served to soak up pressures of urbanization such that their impacts are politically diffuse even if locally their importance can be severe. In such situations, rural planning has been brought onto the political agenda more because of the growing importance of environmentalism at the national scale than any political perception of the needs of rural people.

Planning and policy making have thus evolved for different stated reasons and over different scales of time and space in different nations. Yet these are by no means the only criteria which account for the variegation of planning and policy mechanisms throughout the developed world. As Gordon Cherry (1982: 109) has pointed out:

> planning is grounded in socio-economic, cultural and political contexts; its legitimacy springs from society and is fixed in political and institutional frameworks.

It is therefore reasonable to suppose that an understanding of rural planning and policy making based on the intensive study of *one* socio-economic, cultural and political context and fixed in *one* political and institutional framework would greatly benefit from comparable research in other contexts and within other frameworks.

The breakdown of academic parochialism has been one of the central reasons for the development of the *Journal of Rural Studies* as an international and multidisciplinary forum for rural research. It is this same attack on parochialism which has led to the development of two collections of essays which focus on rural planning from an international perspective. This, the first collection, deals with policies and plans for rural *people*, and the second book will concentrate on rural *land and landscape*. It is freely admitted that this represents a rather artificial division of rural planning matters. Clearly, policies which aim to support agricultural production will impact upon both rural land and rural farm populations. Nevertheless, it is hoped that by bringing together a wide-ranging body of information and reflection from human and environmental perspectives a full appreciation of rural interdependency will emerge at a level which would not be permitted by a single collection of inevitably sketchy overviews of all rural policies and planning mechanisms in a particular nation.

As it is, authors have been presented with a most difficult task of providing informative and incisive commentaries within tight confines of length. This being so, they have been asked to concentrate their approach on specific and common issues: particular problems experienced by rural people, the relationship between rural people and the distribution of political power, the availability of planning and policy-making mechanisms, the implementing agencies concerned, and the nature of central–local state relations. Underlying this common agenda is a wish to illustrate from a wide variety of socio-economic and cultural contexts and political and institutional frameworks:

(a) any communality of problems experienced by particular fractions of the rural population;
(b) any comparability in the planning and policy mechanisms which are developed in response to these problems;
(c) any supportive evidence of contemporary concepts of the rôle of planning within the form, function and apparatus of the state.

These issues are reviewed in the final chapter of this book.

Problems and Rural People

It would be predictable and maybe even expected at this point that the thorny issue of defining 'rural people' should be raised. After all, two clearly diverging schools of thought have emerged on this issue in recent years. There are those who would point out that in most developed nations all people are culturally if not physically urbanized, and that the underlying causes of the social and economic problems faced by people living outside cities are exactly the same as those experienced by urban dwellers; namely, powerlessness, exploitation, uneven distribution, inequality and so on. According to this viewpoint, there is little to be gained from the categorization 'rural' which indeed induces the negative tendency of ascribing spatial explanations to social phenomena. From the opposite corner it is stressed that rural land use, landscape and settlements are patently different from their urban equivalents by dint of scale, density, remoteness, and predominant forms of economic production, especially agriculture. This being so, people who are constrained to live in these areas, or who choose to do so, reflect these inevitable differences in their living environment.

It is not the intention here further to describe or fuel this debate, especially as detailed discussion is available elsewhere (Cloke 1985a, 1988, Cloke & Park 1985). The point surely is not that there is no difference between urban and rural environments, because there patently is, but that the understanding and explanation of social and economic problems encountered in those environments should not be constrained by factors of rurality, marginality, peripherality or any other 'spatiality'. Rather, the very fruitful explanatory areas of social composition, structure and relations should be allied to a view of planning and policy making activities as an integral part of the overall rôle played by the state to form a fundamentally non-rural, indeed aspatial, portfolio of conceptual tools for understanding what goes on in these areas we call rural. Because social relations exist in spatial arenas (Massey 1985) it appears legitimate to make use of pragmatic spatial divisions such as 'rural' to study both the changing composition of society and structuring of the economy as driven by the engine of capitalism, and the activities of the state to regulate these changes through its planning functions.

Part of the chequered history and indeed contemporary state of rural planning and policy making *as presented by its analysts and commentators* stems directly from the fact that different perceptions of rural problems can and do lead to different planning 'responses'. Put bluntly this means that a spatially conceived problem will generate a spatially conceived prescription for response. For example, the socio-economic problems of many remote rural areas in developed countries in the mid 20th century were reflected by trends of depopulation. Following the spatial model, these problems became characterized as 'small town' problems, and indeed a great deal of useful empirical work has been produced which illustrates the changing living conditions experienced by people living in small towns (see, e.g., Swanson *et al.* 1979, Hodge & Qadeer 1983, Johanssen & Fuguitt 1984. However, the perception of the problem has structured the response. The small town problem has tended to be converted into the response of how to provide facilities in small towns (not, more specifically, to *small-towners*).

Throughout the developed world (as illustrated by the following chapters) rural growth centres have been planned at various scales and with varying degrees of success as spatial responses to the social problems in depopulating communities. It is therefore understandable and inevitable that when a reversal of population trends in rural areas is detected as is now the case with the counterurbanization (or counterdepopulation – see Cloke 1985b) phenomenon governments will not only proclaim the success of their planning policies but also, more importantly, withdraw from the need to plan because rural problems have somehow been 'solved'.

As studies of the problems experienced by rural people have become more sophisticated, welfare-oriented concepts of deprivation and disadvantage have assumed greater importance. Even so, these essentially *social* and *economic* problems have tended to have been imbued by the same spatial planning culture as described above. McLaughlin (1986) suggests that deprivation can be explained in two very different ways:

(a)   *A sociological model*, which identifies the root causes of deprivation within the structure of society. From this perspective deprivation is derived from inequalities in the distribution of social status and of political and economic power. Moreover, the influx of adventitious and affluent middle-class newcomers within rural communities has heightened the relative deprivation of non-propertied households, whether relativity is recorded in subjective or objective terms. Social change in turn impacts upon resource allocations of goods and services in both the public and private sectors. Thus social differentiation and stratification are viewed as causes of rural deprivation (Newby *et al.* 1978).

(b) *A planning or services model*, which focuses on the decline in quality and quantity of rural services in rural areas as the major key to the understanding of rural deprivation. The impact of service losses is experienced disproportionately by non-mobile and other disadvantaged groups, although attention has been paid to *place-related* provision of access and service opportunities as the principal planned palliative (Moseley 1979, Shaw 1979).

These categories (along with the work from which they are derived) have been criticized by Lowe *et al.* (1986). Both models are seen as demonstrating 'crucial weaknesses'. In terms of the sociological model:

> The emphasis on social conflict, adopted by Newby and his colleagues as the key to rural power structures, contradicts the idea of a communitarian rural idyll. Even so, the authors of the sociological theory do not assess the rôle of state interventions in contributing to the political and economic subordination of the rural working class. Whilst it would be wrong to criticize this research for not directly addressing the relationship between social deprivation and state action since a study of deprivation was never part of the authors' brief, the lack of attention paid to state mediations of power, property and social control is curious. (pp. 20–1)

The point being made is a simple one, yet the rôle of state intervention on behalf of capital interests as opposed to other interests requires detailed consideration. Some discussion of this point is offered in the next section of this chapter, where questions are broached as to why and how planning and policy interventions take place within the overall functions of the state.

The criticisms by Lowe *et al.* (1986: 21) of the planning model of deprivation reflect similar concerns:

> Shaw and Moseley, in their separate contexts, concentrate on what might be called 'planned deprivation' . . . Attention is focussed,

therefore, on planning as a state activity, of which the distributional impact within the civil society is unexplored. Moreover, questions of the social distribution of needs *within* rural communities, or of the social determinants of political expressions and responses to need, receive scant attention.

Here again, attention is drawn to the distributional effects of resource distribution and market regulation within the context of the state, and clearly a view of rural deprivation derived from political economy approaches will be linked with sociostructural rather than spatial planning 'responses' to the problem.

McLaughlin's main aim in putting forward these two models of deprivation appears to have been to drive home the argument that government policy in Britain has interpreted rural deprivation within the context of *service provision*. He shows that the emergence of rural deprivation as a political issue in the 1970s was, in fact, closely linked to a campaign by some rural local authorities against the urban bias in the allocation of financial support from central government. By correlating rural deprivation with declining rural services and hence with declining spending power enjoyed by rural local authorities, deprivation was being used as a crude but effective political lever. Perhaps more importantly, it clearly directed government responses into a *spatial* rather than *social* realm. McLaughlin (1986: 292) concludes:

> By focusing the problem analyses and subsequent policy prescriptions on the issue of rural *areas* as poor places and on questions of service decline *per se* the policy debate on rural deprivation has largely ignored crucial questions about the particular groups and individuals *within* rural areas who gain or lose as a result of service policies . . . the key issues of differential standards of living and quality of life in rural areas and the resource distribution processes affecting them have also been ruled off the agenda.

There are important implications from this deprivation debate for the wider relationship between rural problems and planning and policy making. Firstly, if we over-readily accept planning policies for rural areas *in terms of their own objectives*, then our understanding of rural planning will be both superficial and sterile. Equally, if we assume that planning and policy activities reflect rational and logical 'responses' to rural problems, there is a grave risk of ignoring the overriding functions and rôles of the state which imbue that part of state activity which we call planning. If deprivation is to be seen in purely spatial planning terms then the evaluations of spatial policies concerning key settlements and accessibility might permit rather optimistic and congratulatory evaluations of rural planning during the 1970s and early 1980s. But as deprivation is now seen by most to have far-reaching derivations in terms of the power relations in state and society, the planning and policy-making activities in rural areas should be evaluated in these state and societal contexts.

What is true for rural deprivation applies to all other problems experienced by rural people. Preconceptions of the rôle of planning within the state context should therefore be carefully acknowledged and understood prior to any evaluation of the success of policies and plans for rural people.

The state, planning and intervention

Peter Hall's description of planning as 'all things to all men' assumes added significance when to the various definitions of planning are added the various conceptualizations of planning context. It would be inappropriate here to offer a detailed discussion of the many contemporary theories of the state. Excellent reviews are available elsewhere from Saunders (1979), Dunleavy (1980), Cooke (1983), Clark & Dear (1984) and Ham & Hill (1984), and a detailed discussion of rural areas in this context has been tackled by Cloke & Little (in press). It is important, however, in the introduction to a series of international essays on planning for rural people to present a brief account of the nature and rôle of planning intervention as seen from different perspectives.

THE NATURE OF PLANNING

Healey *et al.* (1982: 18) offer a succinct list of the various elements of planning which have been commonly recognized in various attempts at definition:

(a) an activity of a particular type (such as rational procedures for the identification and selection of policy alternatives);

(b) an activity undertaken by a particular type of institution, such as government (as opposed to the market);

(c) an activity involving the guidance or regulation of particular classes of events and objects (as in the regulation of land use);

(d) an activity undertaken by people who consider themselves to be planners or to be undertaking planning (the subjective 'planning is what a planner does', and the objective 'planning is what people recognise as what planners do').

Such activities represent a familiar amalgam which commentators and analysts have readily incorporated into their evaluations of the progress and success of planning in rural areas. These defined rôles, however, offer a rather restricted overview of planning unless the relevance of the underlying policy context is also made explicit. Therefore our view of the nature of planning should also seek to take account of the wider aspects of policy planning systems. In a recent review, Hambleton (1986: 144) has produced a series of four propositions about the rôle and operation of policy planning systems:

(1) policy planning systems are intended to promote more 'rational' approaches to public policy making;
(2) policy planning systems are ways of structuring interorganizational relationships and conflicts which lead to the advantage of some parties and to the disadvantage of other parties;
(3) policy planning systems are instruments of central policy control;
(4) policy planning systems are central government instruments for restraining local social spending and/or for channelling spending in favour of economic priorities.

Figure 1.1 demonstrates both the theoretical underpinnings of each of these propositions and the major areas of implementation which they impact upon.

In a discussion of these four propositions in the British context, Hambleton first notes that, although policy planning systems were originally intended to promote rational approaches to public policy

Figure 1.1   Understanding policy planning systems: a conceptual map.

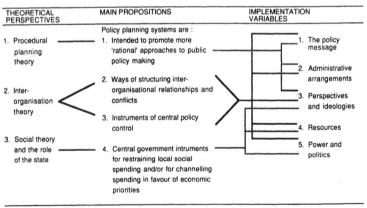

Source: Hambleton (1986: 145).

making, there has been an erosion of this focus because of the increasing awareness of the politicized nature of the planning process. Secondly, he confirms that interorganizational relationships have been structured in favour of central government and the interests they represent, although he advises that care should be taken in this area of analysis as planning systems overall are relatively powerless instruments for tackling entrenched interests within society. Thirdly, Hambleton suggests that policy planning systems have not been used to a great extent as instruments of central policy control. Rather, the central government has chosen to make use of specific sectoral legislation, relating for example to housing, transport and so on, which does a more efficient job of extending central control. Planning has therefore been bypassed in this respect and tends to be viewed by government as a limited but useful channel for central–local negotiation and bargaining. Lastly, the rôle of planning as a restraint on local expenditure is rejected by Hambleton. Again, the formal planning system has been bypassed by more specific legislation and by government practices such as cash limits. The rôle of planning here has often been to direct *where* cuts should be implemented.

Figure 1.1 reflects that the *activities* of planning are responsive to the underlying *policy aims* of the state agencies concerned. In turn, these policy aims will be conditioned by the form and function of the state of which planning and policy making are but one part of the apparatus. Different views of the function of the state will offer different explanations for state intervention through planning and policy making. Three different snapshot views are now presented in order to demonstrate the wide variations in the potential explanation of the state's desired function for planning intervention. It is readily admitted that this is neither a comprehensive nor representative selection. Nevertheless, many of the major discussive issues arise from these views of intervention.

STATE INTERVENTION, THE MARKET AND THE INDIVIDUAL

Many of the more liberal views of planning intervention stem from welfare economics. Willis (1980: 30), for example, describes the underlying principles to this genre of intervention:

> A number of assumptions are necessary for social welfare to be maximised in the free market economy. Where these assumptions do not hold, a pure free market economy is not optimal for achieving this and government intervention is normally suggested.

The most frequently recognized imperfections in a free market economy are that

- the world is not one of universal competition,
- many natural monopolies exist,
- externalities (social costs) exist,
- income inequalities exist,

and intervention has traditionally been justified in terms of the Pareto function, which claims to describe an efficient situation whereby no welfare benefit may be added to an individual or group without some other individual or group losing out.

Planning intervention derived from the desire to overcome market imperfections can, with the eye of faith, be seen to have pursued three separate yet interrelated directions: a rule-making and enforcement policy strategy has been developed to cope with common property or externality problems; a public investment strategy has been pursued to augment capital and resource stocks for collective consumption (for example in roads, housing and so on); and a conservation strategy has been concocted to deal with the conflicting desired use of environmental resources, for instance by developers and ecologists.

If the motive and objective of planning intervention be this kind of regulation of market imperfection, then planning is certainly the subject of political threat in many Western nations where political opinion has slid to the right and capital interests have seized the opportunity once more to hide behind the market as a mechanism of 'fair' resource allocations. In these terms one method of 'improving' planning is to reduce (or at least give the impression of doing so) the imperfections of the market, thereby reducing the need for planning. Thus better planning equals less planning. Such reforms are commonly underwritten by a comparison of market imperfections and political imperfections. For instance, Willis (1980: 257–8) suggests that

> there is a growing tendency in planning to use political rather than market mechanisms to resolve social and individual problems . . . New political power is far more concentrated and unevenly distributed than market power even can be, and there is not a great deal that can be said in favour of it.

Such arguments, although attractive to many, appear to take little account of the confluence of political and economic power which many theoreticians would regard as being enjoyed by various fractions of the capitalist classes. Supporters of the market mechanism purport to maximize the interests of the individual:

> In the economic market, each of us can decide for himself. Within the

limits of our income, we can be sure that what we vote for with our dollars we get . . . That is why the economic market is so far the only mechanism available that provides real individual democracy. (Friedman 1975: 7)

Yet for many rural people 'within the limits of our income' is the most important and constraining factor in this argument. Equally, the logical progression of the more-market-less-planning approach is to tie beneficiaries to costs through a system of charging policies connected to any planning projects undertaken. Willis (1980: 259) suggests that such a policy would 'rupture the linkage that sustains the ability of vested power to dominate public resources and environmental policy through the political process'. An alternative view would be that those who control the market represent a central element of this vested power and that any charging policies would in practice be used further to discriminate between those who can pay and those who cannot. For example, if the cost of conserving small rural settlements in pressured locations was to be charged to the beneficiaries, that is the residents of those settlements, such a policy would merely accelerate the existing trends of social polarization whereby only the affluent gentry can afford to buy housing and other opportunities in these places.

These reforms to the welfare economics basis for planning intervention are currently an increasingly important component of political currency in many Western nations. Cherry (1982: 114) notes the existence of right-wing political proposals based on the assertion that

certain private markets can be effectively self-regulating, without the need for much state supervision or guidance. 'Planning free zones' have therefore been advocated, favouring private markets in health, housing, education, welfare and land.

If the state function dictates this rôle for planning, then specific explanations can be put forward for any attempts to remove or subdue any existing planning machinery for rural areas.

THE STATE ON THE SIDELINES: INTERVENTION IN A MIXED ECONOMY

Another view of the function of the state (thereby encapsulating the function of planning) characterizes the state's activities as protecting and reproducing existing social structures and existing social relations of production (Mandel 1975). Planning intervention in this context would depend very much on whether such protection and reproduction was in some way *aggregate* in nature, and therefore potentially for the benefit of all individuals and groups in society, or whether protection and reproduction were *selective* and beneficial only to specific powerful interests. By accepting the former interpretation (i.e. aggregate protection and reproduction) the intervention rôles performed by the state in terms of planning and policy making can be outlined as (Clark & Dear 1984):

(a)   regulation of the private market *for the common good*;
(b)   *arbitration* of competing bids for resources;
(c)   adjustment of the market according to the state's own *normative* goals, in some form of social engineering;
(d)   guaranteeing the maximum freedom for individuals.

The key concept here is that the state acts as some form of neutral and benevolent arbitrator deciding on the common good, translating this into normative goals and implementing them, in a rational manner so as to maximize individual freedoms. Broadbent (1977) has painted this kind of picture in his analysis of the three *estates* – the state, private (profit) sector firms and households constituting the labour force – in economic planning. He argues that the mixed economy is preserved only by ensuring that the private sector remains profitable and thus in a condition to reinvest:

> For this reason the state does not invade possible profitable areas and tries not to take an entrepreneurial role; it remains on the sidelines, responding to the market – providing grants and incentives to firms (such as in regional policy) – and sometimes tries to put more of the burden of taxation on to the population or to reduce the costs of social services by reducing their quality. (p. 240)

Again, a key question here is whether power relations within the state are indeed as pluralist as these characteristics of neutrality and lack of direct involvement imply. If power is available to all sectors of society through democratic processes, then it can be suggested that the state will not be permitted to generate any consistent bias towards particular classes. Thus planning is the absolute servant of Benthamite democracy, and will only indulge in adjustment or regulation of market trends within the constraints of the need for profit for reinvestment, and within the dictates of majority requirements. Catanese, an American planner, argues this case most strongly:

> Planning is in the service of individuals in the community. When the planner disagrees with the majority of those individuals, he or she should leave that role. Some argue against this assertion and insist that the planner should be an agent of change, even if that means radical and

revolutionary change. That means that the planner is charged with the responsibility for societal change. When the change is an attempt to reorder basic values, however, something is wrong in the concept of public service. The only way out is for someone else to do the planning . . . If the values of a community are so corrupt that constitutional rights or moral order are imperiled then the more appropriate change is through the political or legal process. Unless the planner is directly participating in those processes, he or she should not be trying to refute the beliefs of the community. (Catanese 1984: 39).

Evidently, attitudes such as these reflect a protection and reproduction of societal status quo through planning. A major criticism of this approach, however, has arisen from analyses of the assumed pluralist power relations which underlie the state-on-the-sidelines view. Newby (1979), for example, suggests that in the British example of rural planning there was initially a progressive Fabian desire for reform of the *laissez-faire* approach which had prevailed up to the late 1940s. The intention was to avoid the social injustices of development patterns sponsored by the operation of an unfettered market. But the results of planning intervention based on these objectives have been very different from those anticipated. The planning system designed to meet the needs of the least fortunate has had the opposite effect:

So far it has been the most privileged members of English rural society who have benefited most from the operation of the planning system in rural areas, while the poor and deprived have gained comparatively little. (Newby 1979: 237).

Such outcomes should not be blamed on rural planners *per se*, but both professionals and politicians have failed to ensure that their actions were prescribed by any philosophy of distributive justice.

Similarly, Ambrose (1986), when reviewing planning in postwar Britain, concludes that from the viewpoint of the 'public at large' the achievements of state intervention have been largely tactical and limited in impact. Moreover, although planning alongside the welfare state has exhibited a sometimes paternalistic, sometimes legitimating token compassion for the lower classes, the greatest proportionate benefits from planning intervention have been enjoyed by corporate business enterprises. A planning system which some at least perceived to be able to stand up against capital interests as formulated in the late 1940s has been totally inadequate as a check on the continued process of accumulation driven by much reorganized and restructured units of capitalist production.

INTERVENTION ON BEHALF OF CAPITAL AND CLASS

A different view of state intervention through planning and policy making is offered if the state's function of protection and reproduction is operated selectively in favour of particular interests. Given the widespread critique of pluralist power relations in the state and its planning function, wider credance has been given by many to concepts of state which emphasize:

- *élitism*, where power is vested in minority élite groups within an institutional setting;
- *managerialism*, where power is wielded by professional bureaucrats and gatekeepers;
- *structuralism*, where state power represents the current balance of class interest.

In particular, structuralist views appear to offer a further perspective on planning intervention. If the balance of interests favour monopoly capital, as in Britain, then planning and policy within the state will essentially furnish the needs and interests of capital. Alongside these major objectives it would be expected that short-term and limited reforms would be introduced both as a legitimation of the balance of power and to ensure the long-term interests of the dominant class.

In these terms planning has been viewed alternatively as some kind of parasite with no real rôle in the structures of economic production, or as an instrument of class rule and a system of political domination. Recent analyses stress this latter point:

> At the empirical level there are clear signs that in the U.K. the planning system is passing from an emphasis upon containment and concentration with decentralization, towards a mixture of 'strategies', increasingly involving the regional and local state, and the central state operating in local space. The clearest tendency is for the state to be assisting capital, in whichever ways it can, to escape from the recent reduction in the power asymmetry between capital and labour. (Cooke 1983: 274)

Holistic concepts of rule and domination, however, appear oversimplified in their description of a very complex state phenomenon. Many studies (from Castells 1977 to Cockburn 1977 to Simmie 1981 and many others) have demonstrated that the state is not just a puppet of capital interests and ruling classes. It has its own energy and power; it periodically acts against the interests of particular capitalists in order to support others (Hague 1984); and it consists of a complicated interplay of managers and political and economic interests. A need has therefore been recognized to take account of a state which is simultaneously founded in the social relations of capitalism and capable of generating power and authority of its own (Clark & Dear 1984).

Saunders (1979, 1981) acknowledges this dualism in his categorization of state functions:

(a) *the sustenance of private production and accumulation* by orchestrating demand, restructuring spatial aspects of production, providing infrastructure and so on;
(b) *the reproduction of labour power* through collective consumption;
(c) *the maintenance of social order and cohesion* through political 'participation', coercion, support for economically surplus populations and by the provision of legitimating services such as education and health.

Saunders extends this analysis by suggesting that state intervention occurs at two levels:

(1) The *corporate* level where intervention is implemented so as to favour capital interests. The market regulation of agriculture through social investment policies is an example of this level.
(2) The *competitive* level where services are provided for dependent populations through consumption policies, for example, for housing provision.

Using Saunders's framework, the analysis of planning and policy making for rural people can assume a different character of explanation. Corporate level intervention can be interpreted as being very much wrapped up in the capitalist state and its fundamental support for capital interests. Competitive level intervention represents an arena where some autonomy from capitalism can occur, but only within the bands of constraint established by the needs of intervention at the corporate level. It might therefore be argued that postwar rural planning intervention in some nations such as Britain has been predominantly at the *corporate* level, placing priority on the provision of infrastructure for capital accumulation and on policies favouring agricultural capital. Although there are now increasing levels of conflict between different factions of capital (hence the slow and often reluctant policy changes in agricultural support), the provision of infrastructure for industrial capital remains a high priority for development planning and development control.

At the *competitive* level, intervention has been far less important than control, and this statement of priorities represents a contextual explanation for the apparent failure to secure radical social planning policies in rural areas. Furthermore, in recent years, the willingness of the state to indulge in legitimation policies of social consumption appears to be declining.

Rural planning and policy-making conflicts are often illustrated as a tussle between physical land-use planning and socio-economic planning, with the latter heavily subjugated to the former. It is perhaps more appropriate to characterize these issues in terms of corporate and competitive intervention. Corporate intervention for production has received high priority from the state because of the underlying state function of supporting capital interests and providing an appropriate environment for capital accumulation. Competitive intervention for consumption represents the minimum level required for legitimation purposes, plus any additions derived from the relative autonomy of the state within the bounds set by its relations with capital interests.

The case studies

What follows is a series of case studies which attempt, in the context of particular nations, to evaluate past and present policies for rural people. Each chapter makes an assessment of the problems experienced by such people, the planning mechanisms and policy-making systems which are employed to respond to these problems, and the political and state contexts in which plans and policies are set. All except one of the case studies are drawn from developed Western nations in North America, Europe and Australasia. The rationale here is to sample different sociopolitical contexts within a range of scale and economic development. The exception is Judith Pallot's chapter on the USSR, which is included partly to seize upon the sociopolitical contrasts between the West and Eastern bloc countries, and partly to highlight the fact that, at least superficially, some comparative policies for rural people emerge from contrasting contexts. A more detailed analysis of comparisons and contrasts is presented in the final chapter.

References

Ambrose, P. 1986. *Whatever happened to planning?* London: Methuen.

Broadbent, T. 1977. *Planning and profit in the urban economy*. London: Methuen.

Castells, M. 1977. Towards a political urban sociology. In *Captive cities: studies in the political economy of cities and regions*, M. Harloe (ed.), 61–78 Chichester: Wiley.

Catanese, A.J. 1984. *The politics of planning and development*. Beverly Hills, Calif.: Sage.

Cherry, G. 1982. *The politics of town planning*. Harlow: Longman.

Clark, G. & M. Dear 1984. *State apparatus: structures and language of legitimacy*. Boston: Allen & Unwin.

Cloke, P.J. 1985a. Whither rural studies? *Journal of Rural Studies* **1**, 1–9.

Cloke, P.J. 1985b. Counterurbanisation: a rural perspective. *Geography* **70**, 13–23.

Cloke, P.J. 1988. Rural geography and political economy. In *New models in geography*, R. Peet & N. Thrift (eds), London: Allen & Unwin.

Cloke, P.J. & J.K. Little in press. *The rural state?* Oxford: Oxford University Press.

Cloke, P.J. & C.C. Park 1985. *Rural resource management*. London: Croom Helm.

Cockburn, C. 1977. *The local state*, London: Pluto Press.

Cooke, P. 1983. *Theories of planning and spatial development*. London: Hutchinson.

Dunleavy, P. 1980. *Urban political analysis: the politics of collective consumption*. London: Macmillan.

Friedman, M. 1975. *There's no such thing as a free lunch*. La Salle, Ill.: Open Court.

Hague, C. 1984. *The Development of planning thought*. London: Hutchinson.

Hall, P. 1974. *Urban and regional planning*, Harmondsworth: Penguin.

Ham, C. & M. Hill 1984. *The policy process in the modern capitalist state*. Brighton: Wheatsheaf Press.

Hambleton, R. 1986. *Rethinking policy planning*. Bristol: School for Advanced Urban Studies.

Healey, P., G. McDougall & M.J. Thomas 1982. *Planning theory: prospects for the 1980s*. Oxford: Pergamon.

Hodge, G. & M. Qadeer 1983. *Towns and villages in Canada*. Toronto: Butterworth.

Johanssen, H.E. & G.V. Fuguitt 1984. *The changing rural village in America*. Cambridge: Ballinger.

Lassey, W. 1977. *Planning in rural environments*, New York: McGraw-Hill.

Lowe, P., T. Bradley & S. Wright (eds) 1986. *Deprivation and welfare in rural areas*. Norwich: GeoBooks.

Mandel, E. 1975. *Late capitalism* (translated edition). London: New Left Books.

Massey, D. 1985. New directions in space. In *Social relations and spatial structures*, 9–19 D. Gregory & J. Urry (eds), London: Macmillan.

McLaughlin, B. 1986. The rhetoric and reality of rural deprivation, *Journal of Rural Studies* **2**, 291–308.

Moseley, M.J. 1979, *Accessibility: the rural challenge*. London: Methuen.

Newby, H. 1979. *Green and pleasant land?* Harmondsworth: Penguin.

Newby, H., C. Bell, D. Rose & P. Saunders 1978. *Property, paternalism and power*. London: Hutchinson.

Saunders, P. 1979. *Urban politics: a sociological interpretation*. London: Hutchinson.

Saunders, P. 1981. Community power, urban managerialism and the local state. In *New perspectives in urban change and conflict*, M. Harloe (ed.), 27–49 London: Heinemann.

Shaw, J.M. (ed.) 1979. *Rural deprivation and planning.* Norwich: GeoBooks.

Simmie, J. 1981. *Power, property and corporatism: the political sociology of planning.* London: Macmillan.

Swanson, B.E., R.A. Cohen & E.P. Swanson 1979. *Small towns and small towners.* Beverly Hills, Calif.: Sage.

Willis, K.G. 1980. *The economics of town and country planning.* London: Granada.

2 Britain

PAUL CLOKE

The political climate of rural areas

To describe and assess the policies and plans for rural people in Britain without first understanding the prevailing political allegiances in these areas would be like evaluating the fate of the Titanic without mentioning the iceberg. Planning in Britain is inextricably linked to political decision making at central, regional and local levels, and represents one aspect amongst many in the portfolio of state activities. Analysis of plans and policies therefore necessitates a full investigation of the concepts of state form, function and apparatus (see Ch. 1, and Cloke & Little in press), and an awareness of any variations in state activity due to party political changes.

There has been remarkably little attention given to the rural aspects of national government political trends in Britain. In one of the few specific analyses, Gilg (1984) suggests four main periods of political attitudes to the British countryside:

(1) *The 1945–51 Labour government*
    In this period, stability was granted to farmers under the 1947 Agriculture Act which provided price guarantees and grant-aid for investment. Government also directed postwar planning with the 1947 Town and Country Planning Act which established *development plans* at the county council level and introduced the need for developers to obtain planning permission from local authorities for most forms of development.
(2) *The 1951–64 Conservative government*
    On the fall of the Labour government in 1951, the Conservative administration broadly followed on with these measures, although they were ideologically committed to a planning system which regulated rather than directed rural change.
(3) *The 1964–74 Labour and Conservative governments*
    The 1964 Labour government was marked by a rather pragmatic stance on policy issues and so rural policies reflected a 'more-of-the-same' approach. The successor Conservative government (1970) took a similarly centrist position but was responsible for reinforcing sup-

port for the agricultural sector via entry to the EEC in 1972. This period saw the introduction of the 'new' planning system in 1968 which opted for a broader, more flexible set of county-level policies encased in *structure plans*, with detailed development programming being allocated to the lower-tier districts in the form of *local plans*.

(4)   *The 1974–82 Labour and Conservative governments*
With the world oil crisis and the onset of economic depression, there was a broadly perceived need to relax planning controls and to encourage economic growth. For the first time an anti-farming lobby of some significance showed its hand and began to argue against the positive discrimination granted to the agricultural economy. A major component of this lobby was the 'new' adventitious rural resident whose buying power permitted unrestricted entry into rural housing markets which were often rapidly escalating in price. Planning for rural people began to take the shape of the marketplace as the ethic of intervention became bound by economic rather than social objectives.

In his review of these periods, Gilg (1984: 257) stresses that 'there have been continuities in policy and attempts to provide integration', and certainly it does appear from the main themes of these four periods that variation in national party politics have not been reflected in varying rural policies. There is, however, a danger in accepting this conclusion without reflecting on the functions of the state (as opposed to government *per se*) in rural areas. It could be argued with some validity that throughout the postwar period the state was acting despite potential political conflict to encourage the promotion of a flourishing environment for capital accumulation in agricultural, development, and other industrial sectors. The 'continuity' of policy therefore represents a failure on behalf of Labour administrations to effect radical policy changes in order to intervene against capital and class interests in rural areas which are part of the natural political constituency of the Conservative party. Therefore, all rural policies in the postwar period can be said to be conservative in nature. The accepted political issues have been over the *degree* of conservatism which underlies rural policies. The activities of the Thatcher government have presented a rather more stark and high-profile example of these state functions of supporting capital and class interests, and therefore, rather than accepting 1974–82 as a useful political period, it is more beneficial to look at the beginnings of Thatcherism in 1979 as a marker of a more significant period of political attitudes and activities.

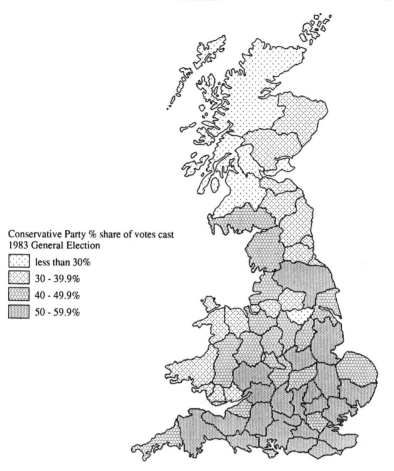

Figure 2.1 Conservative voters in the 1983 General Election, percentages represent proportion of votes cast.

Source: Contemporary reports in *The Times*.

THATCHERISM

At the time of writing, the Thatcher government has enjoyed two terms of office and is approaching its next general election with some confidence. The 1983 election results not only produced a substantial Conservative majority of seats in the House of Commons but also confirmed the Conservative allegiances of almost all rural areas in England (Fig. 2.1). In particular, the prosperous southern rural counties figured significantly as

Table 2.1   Relationships between votes and regions, 1983.

	Conservative	Labour	Alliance	Others	
Scotland	22	45	23	10	100% ($N = 280$)
Wales	34	40	23	3	100% ($N = 167$)
North	39	37	24	0	100% ($N = 812$)
Midlands	52	27	22	0	101% ($N = 573$)
South	53	19	28	0	100% ($N = 1217$)

Source: Heath *et al.* (1985: 75).

strong supporters of the Thatcher ticket. The Conservative vote in the economically more marginal regions (Scotland, Wales and the North) was markedly lower (Table 2.1), and so the two separate nations within Britain began to crystallize. In spatial terms the idea of two nations is descriptive but crude. A better impression of the 'have' and 'have-not' divide is shown by analyses of the relationships between class, neighourhood and vote, such as that in Table 2.2. Using the simple aggregated classifications developed by Heath *et al.* (1985), all classes in 'salaried' neighbourhoods, all classes except working classes in 'mixed' neighbourhoods and salaried classes in 'working-class' neighbourhoods have demonstrated a clear propensity towards voting Conservative rather than for the other two major parties. The two nations concepts should therefore also be interpreted in neighbourhood and class terms as well as within the broader-scale spatial framework. As discussed in the next section, rural areas are typically represented by a high proportion of salaried and other classes than working class, and therefore in all but marginal areas have clear majorities of 'haves' rather than 'have-nots' in Thatcher's two nations.

Rural conservatism in all its forms creates a political environment with two important and interrelated characteristics:

(1)   The majority of rural people have made a conscious decision to live in the countryside and have few complaints regarding the standard of their own life-style and welfare. They are more interested in policies to conserve the landscape in which they live than policies to raise standards of living for the minority of deprived or disadvantaged residents in rural areas.

(2)   The majority of rural people are therefore content with the provisions of Thatcherism and will to various extents continue to lend political support to central government, even if particular policy decisions – such as that to remove the presumption against development on productive agricultural land – are divisive between different class and capital fractions which normally give support to the Conservatives.

Table 2.2 Relationships between votes, class and neighbourhood, 1983.

	Conservative	Labour	Alliance	Others	
salaried individuals					
in salaried wards	55	10	34	1	100% (N = 419)
in mixed wards	60	16	24	0	100% (N = 233)
in working-class wards	48	22	30	0	100% (N = 156)
intermediate-class individuals					
in salaried wards	60	15	25	0	100% (N = 413)
in mixed wards	57	19	23	1	100% (N = 407)
in working-class wards	41	33	26	1	101% (N = 290)
working-class individuals					
in salaried wards	49	23	28	0	100% (N = 177)
in mixed wards	37	40	24	0	101% (N = 264)
in working-class wards	22	61	17	1	101% (N = 435)

Source: Heath *et al.* (1985: 77).

These characteristics are conflated and overgeneralized, but they do help explain the apparent acquiescence of rural people to the policies of the Thatcher government.

Thatcherism in Britain represents a radical political swing away from the previous postwar administrations. It has

> proclaimed the need to modernise the economy, with high public spending as one of the obstacles to the generation, in a competitive market economy, of new, private initiative . . . The new Conservatism combined an attack on what it considered the misapplication of funds . . . with a commitment to promoting greater efficiency through the competition of the market place and reduced public control. While in theory intolerant of excessive levels of state planning and administration, the paradox of the new conservatism was that it had to strengthen the hand of the state in order to implement its policies. (Ryder & Silver 1985: 335–6)

This paradox has important implications for planning and policy making. Firstly, it should be stressed that, unlike many other governments facing fiscal and welfare state crises, central government has assumed increasing levels of power over local government (Goldsmith 1986). It has even moved to abolish metropolitan councils whose Labour-controlled administrations and policies represented a threat to central power. In particular, over the Thatcher years there have been strong controls over local government expenditure which have limited the discretion and autonomy available to local-level planners and policy-makers. Secondly, despite these trends of centralization, local government and not central government continues to bear most of the responsibility for service delivery in most key areas of the welfare state, including housing, education and social services. Service provision agencies in the public sector are thus becoming further removed from the centre of political power, and central government policies of public expenditure reduction, privatization of state services, and an effective reduction in the scope of the welfare state place Britain in a high-profile, right-wing political context.

Local government in rural areas of Britain is solidly under Conservative control in the heartland counties of the South Midlands, East Anglia and Southern England (Fig. 2.2). Although in some county councils an upsurge of Alliance (the middle-of-the-road party) support in 1983 has resulted in a loss of overall control by the Conservatives, it remains the case that it is only those counties on the margins (for example, in Scotland and Wales) and those influenced by urban constituencies in the north of England (for example, Lancashire and Cumbria) where any potential policies for local rural areas might arise from non-Conservative councils. Even so, with increasing trends of centralization, local governments not in Conservative control are heavily constrained to conform to central government requirements in matters of finance, service provision and local policy making. In counties under Conservative control the acceptance of Thatcherite policies is becoming ever stronger, particularly because the old-style Conservative politicians in such authorities (often drawn largely from the farming and landowning fraternities) are being replaced by the new Town Tories drawn from the adventitious in-migrant service classes.

Any understanding of policies and plans for rural people in Britain should be set against this specific background of political power relations.

Changes and problems in rural areas

Changes in rural society and resultant problems have been fully discussed in a number of recent books (Pacione 1984, Phillips & Williams 1984, Cloke & Park 1985, Gilg 1985a, Lowe et al. 1986), and further detailed analysis would be inappropriate here. It is necessary, however, to place the ensuing discussions of rural policies in the broad context of rural problems

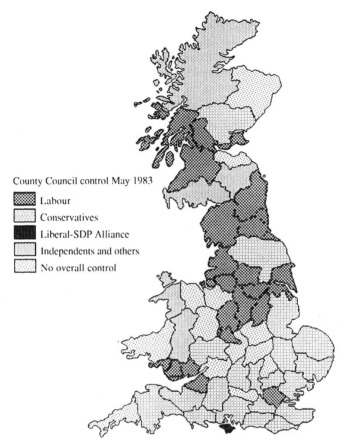

Figure 2.2 Political control of county councils, 1983.
Source: The municipal yearbook.

in Britain. Briefly, the key themes here are counterurbanization, social polarization, resource rationalization and rural deprivation, although the interrelationships between these trends are as important as the individual components.

Counterurbanization was highlighted in the 1981 Census which showed a population upheaval in remote rural areas which had previously been characterized by trends of depopulation (Champion 1981, Robert & Randolph 1983). Table 2.3 shows Champion's breakdown of census data into five rural clusters which differentiate between districts in Scotland and

Table 2.3   Population change in rural local authorities, 1961–81, by cluster type

Cluster	Number of districts	1981 population 000s	Percentage population change 1961–71	1971–81	Difference
7 Rural Wales and Scottish Islands	16	645	−0.2	7.0	7.2
8 Rural, mainly West	32	2009	7.2	8.8	1.6
9 Rural, mainly East	31	2411	15.0	12.7	−2.3
10 Rural, mainly Scotland	23	911	−1.9	9.3	11.1
Subtotal	102	6056	7.5	10.2	2.7
Two rural growth areas	31	2872	22.0	8.6	−13.4
Total	133	8928	11.8	9.7	−2.1

Source: Champion (1981: 21).

Wales (remote), in West and East England (less remote) and peri-urban England (least remote). This analysis uncovers a population revival in the remoter areas, a broad continuity of growth in the less remote areas, and a failure by the least remote areas to maintain previous growth levels.

This population turnaround has been linked with analyses of urban-to-rural industrial shifts (Keeble 1980, Fothergill & Gudgin 1982, 1983, Massey & Meegan 1982). Rural areas in general (although 'rural' here has been at times carelessly defined) have been shown to be the principal recipients of new manufacturing growth during the 1970s, in distinct contrast to the decline in manufacturing which has occurred in inner metropolitan areas. There are also other factors involved in the rural population turnaround (Cloke 1985), including wider issues of retirement, unemployment and long-distance commuting, as well as changes in other industrial sectors, particularly service employment and self-employment.

These changes may profitably be explained by stressing the centrality of capital accumulation as the engine of social formation. In the search for surplus value, capital units will generate unbalanced trends of growth and decline and will restructure production so as to overcome any barriers to accumulation. Restructuring often involves a relocation of production to currently 'favourable' environments, and during the 1970s some rural areas represented such favourable environments for accumulation because of inexpensive and compliant labour forces and localized state subsidies (Bradley & Lowe 1984). By contrast to capital mobility, labour has remained relatively immobile and so recomposition of local civil society has taken place as restructuring has led to societal conflict. Rees (1984: 27) has stressed that

> Changes in rural employment structures are central to any understanding of the reality of rural social life . . . Employment changes themselves have resulted in radical developments in terms of rural class structures, gender divisions, the forms of political conflict occurring in rural areas and, indeed, of the complex processes by which 'rural cultures' are produced and reproduced.

If the irresolvable conflict between capital and labour is to be viewed as central to rural change, as analysis in the political economy genre would suggest, then the definition of planning objectives for rural people, and indeed the identification of the problematic (Cloke 1988), have to be reconsidered. As Healey (1982: 187) has pointed out:

> The notion of irresolvable conflict thus undermines many concepts dear to planners, such as the 'general interest' or a 'balanced strategy'.

It is this dilemma which underlies discussions of 'problems' experienced by rural people. On the one hand, the majority of rural people might be described as relatively 'problem-free' in that increasingly they have chosen to live in rural settlements and are sufficiently wealthy to purchase their life-style without any regulation by planning of housing, service or labour markets. On the other hand, there remains both a remnant of the former agricultural basis of rural society, who are trapped in the locality due to low-paid employment, old age and lack of resources (Newby 1981), and the *nouveaux pauvres*, whose decision to live in those parts of the countryside where housing and land markets permit enforces a life of relative poverty due to lack of earning power. The result is *social polarization* in many rural areas, with planning restrictions on new housing development in small settlements leading to those settlements becoming residentially desirable, to an escalation of house prices, and to a rapid gentrification of local society.

Alongside the gentry, *deprivation* exists amongst the remnant and *nouveaux pauvres* fractions of communities. McLaughlin's (1986) study of deprivation in five case study areas in England recorded 25% of households as living in or on the margins of poverty.[1] Moreover, similar proportions of deprived groups were found in study areas in the thriving south of the country as in the more depressed north. McLaughlin's (1986: 294) conclusions are an important contribution to the debate over social polarization:

Apart from the stark contrast which such findings provide for popular images of affluence which characterise rural areas, they also illustrate the extent to which the rural poor have become statistically marginalised. In contrast to the situation in the historic past where poverty is identified as a shared experience for the majority of the rural population . . . in the late 20th century the rural poor occupy the position of a minority group (albeit a substantial minority).

Should planning and policy making then be directed towards the needs and objectives of the middle-class majority? If so, policies for conservation of settlements would be paramount and planning intervention would only be necessary in a negative mode. Or should policies be geared towards the disadvantaged minority, either in terms of individual subsidies through benefit or taxation systems or by the spatial provision of housing and service opportunities in the places where they live? To a large extent the crux of the matter of rural policies for the minority has been sidestepped because of the high profile given to the problem of the rationalization of resources in rural areas. The loss of services in both the public and private sectors, and the need for rural planning to develop strategies of opportunity provision throughout the hierarchy of rural settlements, has induced a definite *spatial* perspective to policies. By centralizing service delivery into rural growth points and protecting smaller settlements from overdevelopment, the needs of all rural people can be seen to be responded to. Incidentally (or as a matter of political power relations) any prospect of social rather than spatial policies for the deprived is rendered unnecessary by this emphasis. Spatial policies can be demonstrated to be for the good of all rural society, whereas specific social policies for the deprived would be divisive and politically unacceptable in rural Britain, given the political context described above.

Formal policies for farm populations

Thirty years of agricultural expansion – which had brought great prosperity to a few farmers, a better standard of living for many, and for the small and marginal farmers continued hardship and often bankruptcy – came to an abrupt halt in the mid-1980s. A disenchanted public, a critical Treasury and a self-destructive farming lobby conspired to bring about an agricultural revolution which imaginative farmers were already presaging and economists and environmentalists were pleading for. (Pye-Smith & North 1984: 128).

At the end of their book *Working the land*, Pye-Smith and North present an analysis of British agriculture as they predict that it will look in the year

2000. Central to this prognosis is that the mid-1980s will represent an important watershed in policies for agriculture. This prediction has proved a sound one.

Government policy for rural areas in postwar Britain has been dominated by the construction of a political system of agricultural support. The drive for increasing productivity, technical sophistication and economic efficiency in agriculture has created a cushioned environment for many farmers. A fourfold increase in production has resulted, based on a transformation of farming from old-fashioned arcadia to agribusiness of a rationalized and capital-intensive nature (Edwards & Rogers 1974, Beresford 1975, Newby 1979).

Newby (1979) details two major objectives which underlie state intervention in agriculture:

(1) to maintain income levels of farmers so that potential out-migration from the agricultural industry could be regulated;
(2) to maintain the stability and efficiency of agriculture.

He highlights the internal contradictions which have beset policies based on these objectives:

> Securing the low-income problems of farmers meant essentially keeping marginal producers in business, yet the whole logic of the other aim of increasing efficiency was to drive them out of business in order to promote amalgamations and economies of scale . . . to redirect money to those farmers who most needed it would be to perpetuate the uneconomic and the inefficient. This was and is the classic dilemma of state support. (Newby 1979: 112).

Nevertheless, the 1947 Agriculture Act established a system of deficiency payments by which farmers received a guaranteed price for their commodities at a level which ensured their profitability, and after entry into the EEC similar aims of state support were pursued through direct intervention in the market.

Thus, for nearly 40 years in postwar Britain central government policies have been committed to the upkeep of the particular fractions of the rural population who were involved in the continuing process of agricultural production. In so doing, certain interests were favoured while others in the countryside were left to fend for themselves in an unprotected rural marketplace. There have been clear 'losers' amongst the rural population because of state policies for agriculture. Perhaps most importantly, the wholesale capital substitution of labour has displaced the employment for agricultural workers which had previously been indispensable for the economic *raison d'être* of rural settlements and communities. Loss of farm jobs sponsored the trends of out-migration and depopulation which characterized many rural areas in the 1960s and 1970s. Some former farm workers (particularly the more elderly) did not move out, however, and have since constituted a major element of the contemporary rural deprived. Even those farm workers who have clung onto their jobs have been losers in some respects. Although they have escaped unemployment, their wages have lagged behind those in other economic sectors. Newby (1981: 231) regards these groups as 'a residual population the flotsam of agricultural change'. Government policies have sponsored radical changes in the composition of rural society without compensatory intervention on behalf of this residual, whose needs have not been considered important and whose plight has been viewed as a short-term issue in the inevitable transition period between the old farm-based rural society and the new market-oriented conditions for rural living.

The other main losers from central government agricultural policies have been within the farming fraternity itself. A combination of underwritten production and incentives for capital investment, particularly in mechanization, has forced farmers into agricultural systems where capital accumulation and investment are paramount. Small farmers have been unable to compete with their bigger brothers in these respects and have suffered accordingly. Blunden & Curry (1985) highlight the plight of the upland livestock farmers who are less capitalized than the cereal barons, whose products have not been central to the state's price support measures, who are susceptible to EEC measures such as quota systems, and who are equally susceptible to the potential dangers of bigger lowland farmers switching to livestock production if quotas are placed on arable production. Moreover, the National Farmers Union, which has been so successful in presenting the farmers' case to government, has tended to present the interests of large rather than small farmers. Indeed, it has been acquiescent in the formation of agricultural policies which have squeezed small farmers out of production.

The distributional effects of agricultural policy have largely been ignored by central government policy-makers, who have peddled the myth that a successful and productive agricultural industry benefits everyone and ensures a happy and a healthy countryside. In reviewing the different groups which together constitute deprivation and disadvantage in rural areas, Newby (1981: 226) notes:

It is possible to trace a chain of events from the political economy of modern agriculture to the problems encountered by each of these groups, yet the formulation of agricultural policy has mostly ignored the possible external consequences and at best been indifferent to them.

The agricultural 'revolution' of the mid-1980s, signalled by Pye-Smith and North among others, has certainly not occurred because of a political change of heart over the distributional impacts of previous policies. It is a powerful combination of concern over the potentially calamitous effects of overproduction on EEC agricultural support policies and the realization that up to 20% of Britain's farmland could be redundant given current levels of agricultural production (Centre for Agricultural Strategy 1986) that has precipitated policy changes in early 1987. One of the major proposed changes has been a relaxation of planning controls in the countryside. The existing presumption against development on agricultural land is now to be withdrawn, except for the top two grades of agricultural land and for environmentally élite areas such as national parks. The ramifications of any such policy move will depend very much on central government's own advice to its planning inspectors. If development proposals do accelerate on previously protected land, and if local planning authorities resist these proposals, then it will be when developers take recourse to the *appeal* procedure when government inspectors will rule whether a surge of development is to be permitted. Clearly, however, the domination enjoyed by agricultural capital over the planning and policy systems operating in rural areas has been dealt a blow by these proposals.

As a balance to these planning control proposals, the Minister of Agriculture, Michael Jopling, announced in early 1987 a £25 million package of proposals designed to buttress the rural economy. Recognizing the need to help farmers to diversify their economic interests because subsidized agriculture will no longer provide the unfailing support that it has in the past, the Minister has proposed a number of policies under the codename ALURE (Alternative Land Use and the Rural Economy):

- £10 million to promote farm woodlands, offering payments to farmers for new planting, much of which will be of coniferous species;
- £3 million to promote traditional forestry on farms;
- £7 million to double the number of Environmentally Sensitive Areas to 18, wherein farmers are subsidized to undertake appropriate farming practices;
- £5 million for economic diversification and marketing.

For farmers, many of whom are for the first time threatened by the partial removal of state 'featherbedding' policies, the 'compensation' offered by these ALURE proposals represents a drop in the ocean, and indeed Michael Jopling was given an unprecedented vote of censure by the National Farmers Union in early 1987. These proposals have a long way to run, but once again it can be seen that agricultural policies, designed for the farm population, have secured the priorities in central government's policy making for rural areas. It is this underlying policy bias which helps to make sense of other formal planning policies for rural people in Britain.

Statutory planning policies for rural people

The formal planning system established in 1947 was built around the progressive political aims of helping the less well-off, but its impact in rural areas has not reflected these aims. Newby (1981: 228–9) explains this paradox:

> The traditional English reverence for the rural way of life has ensured that precisely *what* it was that was being preserved has never been examined too closely. There has been a fallacious belief that the 'traditional rural way of life' was beneficial to *all* rural inhabitants, an influential but unexamined assumption which was the product of an unholy alliance between the farmers and landowners who politically controlled rural England and the radical middle-class reformers who formulated the post-war legislation.

Since the enabling legislation in 1947, formal planning in rural areas has been undertaken by local authorities and has incorporated a strategy of resource concentration into growth centres (Cloke 1979, 1983). The original adoption and continued popularity of these policies stems from a political perception of the economies of resource allocation during periods of financial restriction. Accordingly, the presumption that resource concentration strategies will achieve economies of scale because 'bigger will be cheaper' has proved immensely attractive in policy-making terms. Ayton (1980) has laid down four tenets of economic reasoning which follow the assumption of economies of scale:

(1) small villages cannot independently support education, health and commercial service which require support populations of thousands;
(2) public sector service options are constrained by limited and diminishing resources;
(3) private sector service and some public sector services (for example gas) will not be provided where they are unprofitable, and rural areas often fall within this category;
(4) mobile services incur high running costs and offer a low quality of service.

These four principles lead directly to the practice of fixed-point service provision in sizeable centres. Yet planning intervention to establish and build up these centres (or *key settlements*) has equally been brought about by a pragmatic need to adopt a high-profile policy with which to demonstrate that rural problems were being tackled, and the key settlement policy of resource concentration presented itself as a convenient blueprint, which was both visible and seemingly cost effective.

It is hardly surprising, therefore, that rural policies within the 1947–68 development plan era differed only according to the degree of resource concentration, rather than whether concentration was a more equitable or beneficial strategic option than, for example, resource dispersal. Details of the actual policies are given elsewhere (Woodruffe 1976, Cloke 1983), but in summary three main categories of policy may be identified:

(1) *Key settlement policies* where the comprehensive concentration of housing services and employment into selected centres is sought not only to build up the centres themselves but also to provide opportunities for hinterland villages.

(2) *Planned decline policies* in which direct attempts are made to rationalize the rural settlement pattern by refusing to locate public investment in outmoded small villages and by prompting a population shift into larger growth centres.

(3) *Village classification policies* where villages are categorized according to existing service functions and environmental quality, so that growth can be allocated to suitable (usually larger) receptor settlements.

Some positive achievements of these policies should be recorded. Firstly, in terms of land use, it is clear that policies of resource concentration have usually been effective in preventing sporadic development in the countryside (Working Party of Rural Settlement Policies 1979). By channelling growth into selected centres, key settlement policies have also to some extent aided the increased provision of infrastructural services such as sewerage networks, electricity and telephones. These successes should not be underrated as it is clear that in many areas the concentration and conservation ethic has limited undue urbanization in the countryside and environmental quality in many small villages and has achieved pragmatism in the provision of statutory services.

The weakness of resource concentration policies is to be found in their preoccupation with physical planning to the detriment of social conditions in rural communities. It should be remembered that two basic objectives were sought through these policies – the build-up of centres of opportunity (the key settlements) and the use of these opportunities to improve conditions for residents in hinterland villages.

Some success has been gained in the build-up of key settlements. Case studies in two English counties (Cloke 1979) suggest that concentrated housing and employment provision (the latter often using industrial estates and advance factories to attract entrepreneurs) has occurred in some places and has ensured that rural people do receive opportunities to live and work in the countryside rather than being forced to migrate to higher-order urban centres. However, the second key settlement objective of maintaining villages in the rural hinterland has been less successful than the first. There is clear evidence that the use of resource concentration policies has coincided with a general deterioration of service, housing and employment opportunities in small villages. Although it is difficult to assess the degree to which planning policies are responsible for these trends, the planners' reluctance to permit housing and employment development in non-selected villages has exacerbated rural problems in small settlements. Moreover, there are small but significant numbers of deprived households who have become trapped in these unsupported settlements, as *in situ* services disappear and public transport links to the nearest service centres are not maintained.

Two broad criticisms of the social rôle of key settlement policies appear valid. Firstly, the county-level plans have assumed that there is a standard type of rural community and have ignored the local-scale variations which demand flexible planning solutions. Secondly, these plans have not proved easy to adopt in the light of changing circumstances in rural areas. In particular, rural policies have ignored the scope for the dispersed provision of small-scale housing, service and employment schemes in hinterland villages.

More importantly, a series of problems with resource concentration policies occur outside the remit of current planning powers. Direct provision of rural employment, suitable housing for local rural needs, service opportunities for non-mobile groups, and adequate public transport links between key settlements and hinterland villages are beyond the direct control of planning authorities, yet it is the lack of these various opportunities which has prevented the framework policy of resource concentration from fulfilling its full potential. Isolated cases of positive planning have occurred, but generally there have been insurmountable financial and administrative barriers preventing a co-ordinated approach to rural planning during the development plan era – a situation which Green (1971) has described as 20 years of wasted opportunity for positive rural planning.

The 1968 Town and Country Planning Act required county councils to present *structure plans* and district councils to present *local plans* – the former as a broad-based and flexible programme of strategy for the wide area and the latter as detailed schedules for small-scale development control. Structure plans were to consist of very detailed surveys of country-wide social and economic trends, followed by a written statement of policy intent.

In terms of rural strategies there has been no apparent policy redirection away from resource concentration and towards resource dispersal during the presentation and implementation of structure plans in Britain (Derounian 1980, Cloke & Shaw 1983). The previous criticism of key settlement policies has been ignored by decision-makers in favour of a continuing allegiance to the perceived qualities of channelling growth into rationalized centres rather than dispersing opportunities into the small, more needy, settlements. The reality of the situation for most counties is that they have been faced with insurmountable previous commitments to the development plan framework of key settlements. Planning permissions had been granted for major housing developments, and such permissions are extremely expensive to rescind. Long-term investment strategies based on a centralization of resources had been entered into by agencies dealing, for example, with water, health and education services. These, too, would be extremely difficult to halt and redirect into more dispersal-oriented rural locations. Perhaps most important of all, central government showed no sign of wavering from its tacit support of the economies of resource concentration. Thus, when structure plans were presented to the Secretary of State for the Environment for his approval, such approval was not forthcoming unless certain 'standards' of financial housekeeping were maintained. These standards were such that resource concentration was an inevitable conclusion.

Only in a small minority of counties has there been any apparent attempt to implement policies of resource dispersal. For example, the North Yorkshire Plan (1980) sought a redistribution of opportunity and development in rural areas of the county by identifying socio-economically cohesive village groups which would act as service centres for the less prosperous areas. Opposition to this overt form of resource dispersal was encountered both from various agencies responsible for rural service provision and from the more prosperous districts who objected to the uneven distribution of potential investment. The Secretary of State was therefore able to veto the identified village groups and thus removed the teeth from a potentially innovative policy. The plan for Gloucestershire (1980) has encountered similar difficulties. It, too, supported the idea of village clusters as centres of investment in areas of rural decline, but this intention was also thwarted by resistance at both central government and local government levels.

The story of structure plan policies for rural areas, therefore, is largely of a direct follow on from the development plan penchant for resource concentration policies. Those few attempts to deviate significantly from this theme have been frustrated by resistance, particularly from the central state. Moreover, in many of its aspects, the debate over the relative merits of resource concentration and resource dispersal policies has become sterile in its contribution to an understanding of why rural problems remain unsolved. Two main reasons may be advanced to account for this situation. Firstly, there has been a widespread inability to uncover indisputable links between planning policies and socio-economic circumstances in rural areas. The tendency has been to infer relationships between key settlement policies and various positive and negative outcomes, but these inferential explanations have proved universally unsatisfactory. The position is that the exact impact of resource concentration policies is unknown, despite their use for more than three decades. It is interesting to surmise that the rural socio-economic situation in Britain today may have been arrived at *whether these plans were in operation or not*, such is the strength of economic market tendencies of concentration and rationalization.

Secondly, it has become generally recognized that the strategic level considerations of key settlements and other types of policy cannot of themselves solve rural problems. Policies of resource concentration and resource dispersal merely act as spatial umbrella policies within which the essential acts of resource allocation are performed. Thus the provision of housing for local needs, employment, services and access opportunities is a specific task which requires lower-level and specific decision making within the umbrella planning framework. In one sense, then, to focus rural planning debate on the framework policies is to miss the very important point that most impacts of rural planning occur at the local level as a result of specific decision making. This decision making may not be local, however, as many such local impacts are caused by regional, national and international policy-makers.

Community and development policies for rural people

Given these constraints (see below) on statutory planning activity in rural areas, planners have been driven to seek alternative channels through which intervention on behalf of disadvantaged rural groups might be organized. Essentially this has involved a soul-searching review of available mechanisms of implementation within a political and economic climate which militates against social planning and 'unnecessary' public expenditure.

In a survey of senior planners working in non-metropolitan county planning departments (Cloke & Little 1986), several clear trends emerged as to the available options for implementing policies for rural people. When asked about the major problems of implementing statutory policies, planners reeled off a list of issues.

(a) *interorganizational conflicts* between the county planning authority and other decision-making agencies;

(b) *lack of finance*, particularly because of inadequate global expenditure levels for services, housing and other investment in rural areas;

(c) *lack of control over private sector interests*, a factor which assumes crucial importance where market forces conflict with planning objectives;

(d) *local political and public commitment*, particularly the lack of political resolve for intervention on behalf of disadvantaged groups, founded upon the desire of the middle-class 'floral majority' to conserve the environment which they have bought their way into;

(e) *central government policies*, which have been fragmentary in the planning sphere, and which have overruled local planning considerations in setting out specific financial and organizational policies for public sector investment in, for example, schooling and housing.

The major issue which underlies these problems and which has often been used to legitimize the lack of intervention activity is the extent to which statutory policies within development plans and structure plans should be used for social planning purposes. With the collusion of central government many Conservative local authorities have chosen to downplay social, community and educational policies because statutory plans should mostly be concerned with *land-use* matters.

When asked how these problems of policy implementation might be overcome, practising planners reveal a climate of action which involves using any (usually unconventional) resource which comes to hand in order to further planning objectives. With declining governmental resources, this *planning by opportunism* has largely involved the harnessing of community self-help and voluntary initiatives as part of the planning strategies employed by local authorities for rural areas. For example, the implementation of the Gloucestershire structure plan (Cloke & Little, 1987a, b) focused on a so-called Rural Community Action Project as the main vehicle for intervention on behalf of the rural disadvantaged. The Project looked to forge a partnership between local authorities and rural people in order to produce action. It was, however, scheduled within a pragmatic view of the restricted powers and expenditure available to such an initiative, and therefore it was the local people who were expected to play the central rôle in this partnership. The programme of action connected with the Project (Table 2.4) should therefore be evaluated in the light of this balance of expectation.

The community self-help approach to planning for rural people is double-edged. It is legitimately argued that self-help, supported when required by advice, information and even pump-priming funds from

Table 2.4   Programme for Gloucestershire's Rural Community Action Project.

1   Define project aims and agree draft programme with Chief Executive's office.
2   Discuss and agree aims and programme with Chief Executive and Planning Officer of Cotswolds District Council.
3   Draw up a shortlist of areas suitable for initiating the project.
4   Present project brief and shortlist of areas to Cotswolds District Planning and Development Committee, and full Council.
5   Invite outside service agencies (water, health, post office, etc.) to co-operate in the project.
6   Brief the nominated officers of County Council and District Council service departments whose involvement may be necessary.
7   Brief County Council and District Council elected members for the selected project area.
8   Arrange a joint meeting of elected members, parish council chairman and individuals of local standing in the project area, together with the Community Council, to discuss the project and obtain local approval for going ahead.
9   Design and arrange local publicity.
10 Hold a public local meeting in each of the selected parishes, attended by members and the nominated officers of the service departments and agencies involved. The purpose of the meeting will be to introduce the aims of the project, and invite debate from the floor concerning issues of local concern. Stress the aim that the project is a co-operative venture, with the community providing the lead. Propose a vote for or against the continuation of the project. The meeting to propose nominations for a community group to be responsible for continuing the project.
11 Issues raised in the first public meeting will form the agenda for the first meeting of the local group. Officers of those departments whose responsibilities overlap with the issues raised on the agenda will be requested to attend the meeting.
12 Publish a newsletter following the first meeting, reporting the debate and consequent action. Continue the newsletter thereafter.
13 Officers of the County Council or District Council to provide administrative assistance and expertise until the project appears to be self-sustaining. Subsequently to attend as required.

Source: Gloucestershire County Council and Cotswold District Council (1981).

government agencies, is the most sensitive way to find solutions to local social need in rural areas. Yet in the Gloucestershire case, and indeed more widely, local politicians were unwilling to sanction undue *direct* expenditure or *direct* intervention by local authorities themselves. In this situation, self-help is the only approach which can conform to these political restrictions (which are imposed by the central state as well as the local state) yet still give the impression of a willingness to act in response to the problems in declining rural areas. Self-help could then be viewed as an opportunistic clutching at straws in the absence of any other implementation mechanism. A major change of political will would be required if public sector planning intervention were to become better funded and policy-enabled. Yet there are insufficient voters in rural areas to influence the political will for change. In these circumstances, contemporary policy and action often constitute a political gesture which becomes a sham substitute for resolving actual problems. (Cloke & Little 1986: 283)

There is one important area in which central government has attempted to provide funding and initiatives within the overall climate of planning by opportunism, and that is through the establishment of three rural development agencies. A government quango, the Development Commission, has responsibility for rural socio-economic development. It not only funds the Rural Community Councils, which have now been established in each county with the task of co-ordinating voluntary initiatives in rural areas, and CoSIRA,[2] which is specifically concerned with establishing small-scale industries in rural areas, but it is also responsible for a series of Rural Development Programmes in specific problem areas in rural England (Fig. 2.3) (Development Commission 1984, 1985). Rural development programmes consist of:

(a) the generation of a long-term strategy of responding to social needs, including an evaluation of housing, employment, services and facilities;

(b) the generation of a detailed programme of action, including funding of advance factories and workshops.

These programmes have been recently reviewed by Smart (1987: 210):

With guidance as necessary from the (Development) Commission, they now contain creditable assessments of needs, statements of objectives and self-generating programmes which are monitored and rolled forward each year. Indeed, focusing on economic development, housing, transport, services and social facilities (sometimes even education) and with regular involvement of bodies such as MAFF[3] and the Tourist Boards, the RDPs could in time begin to 'read across' at central government level. Thus they could enable government departments to identify policy and organizational decision issues that might not otherwise be seen.

Despite this potential, the Development Commission currently plays the rôle of a minnow alongside the whale which is the Ministry of Agriculture in Britain's rural areas. Without a major redistribution of financial resources, its activities will of necessity take the form of nibbling at the edge of major structural problems rather than biting deeply into them. Even the recent ALURE proposals (see above), which promote economic diversification

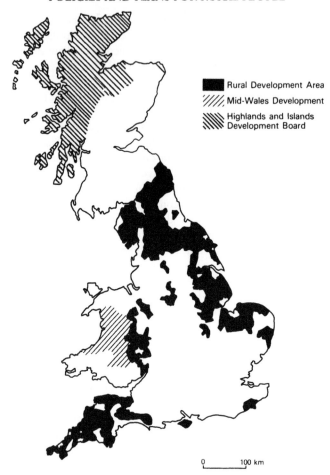

Figure 2.3   Rural development agencies and areas.
  Source: after Gilg (1985: 22).

for farmers and therefore throw a spotlight on existing and future
Development Commission activities, do not carry with them the sizeable
financial investment required to effect substantial changes to the opportunities of the majority of needy rural people.

  The Development Commission's activities are confined to rural
England. In Scotland and Wales, partly as a response to the debates over
nationalism and devolution, specific and separate rural development
agencies have been established. The Highlands and Islands Development

Board (HIDB) (Fig. 2.3) was established in 1965 to promote economic and social development in the northern parts of rural Scotland, and the Development Board for Rural Wales (now Mid Wales Development – MWD) was set up in 1977 to do a broadly similar job. The major task undertaken by both agencies has been the creation of employment, through direct provision of sites and buildings, through grant and loan support for fledgeling enterprises, and through a range of support and advisory services. On a smaller scale both agencies have been enabled to offer grants and loans for the provision of local social and cultural infrastructures. In Mid Wales this has ranged from providing new communal television aerial systems in areas of poor reception to staging concerts by big-name rock bands in obscure locations in the middle of rural west Wales! In an excellent review of these organizations Williams (1984: 82) concludes:

> In an economic sense, the achievement of industrial diversification through the various advance factory development programmes, and the wide ranging training and marketing initiatives available from specialist agencies, has undoubtedly stimulated local entrepreneurial development . . . In community development terms, support for community initiated and implemented social projects in rural Wales and Scotland improves the quality of life in the short term, and may encourage the development of local skills, and motivation for enhanced participation in the long term.

The rural development boards in Celtic Britain therefore offer perhaps the best example of direct intervention on behalf of rural people. The activities and priorities of the boards have certainly not been exempt from criticism, particularly from those who seek specific sociocultural initiatives in Welsh- and Gaelic-speaking areas (see, e.g., Wenger 1980). Nevertheless, with combined annual budgets of nearly £50 million and with specific legislative backing for promotional activity and socio-economic planning, these agencies have suffered fewer of the constraints imposed by central government on their local authority counterparts. Two important issues in a broader political context are noteworthy here. Firstly, despite the success of the boards, the concept behind them has *not* been extended to cover the whole of rural Britain, or even all the remoter areas of Britain. The Development Commission has far fewer powers and resources to employ in England than have the HIDB and MWD in Scotland and Wales. Political legitimation for the Celtic regions is obviously, therefore, an important factor behind their special treatment in this instance. Secondly, HIDB and MWD themselves are under increasing pressure because of Thatcherite policies at central government level. MWD, for example, has already had to change its strategy consequent on the loss of central government regional aid from its area. In many different ways, these boards are having to change their rôle in response not only to the changing expectations of government but also to the veiled threat of resource cuts or even termination of the agencies in their current form. Thus Pettigrew's (1987) recent account of industrial development in Mid Wales reflects the shift from *grant-aid* to *marketing-aid* in line with what he describes as the 'political roller-coaster'. It could be that the high profile and in most respects successful interventions by these public sector agencies will in the end be their downfall in a political régime which increasingly seems to favour the market rather than the planning policy.

Policy, planning and implementation

Rural Britain illustrates paradoxes relating both to the definition of 'what is a problem for rural people' and to the prescription of policy responses to these problems. During the 1970s and early 1980s rural researchers in Britain appeared to establish overwhelming evidence that policy changes are necessary because of progressive imbalances in resource distribution and social equity amongst rural people. Such a view would not, however, be shared by what is now a Floral Majority of adventitious middle-class rural residents whose wealth and income present them with the means to overcome, and often not even notice, the problems experienced by the agricultural remnant and *nouveaux pauvres* fraction of the rural population described above. Equally, it would not be shared by local politicians who have tended to set about their policy-making duties in an entrenchedly conservative manner. Although there are some signs that through motives of paternalism, beneficence, philanthropy or whatever, old-fashioned country Conservatives might be induced into some small-scale policy changes, these tendencies have been unceremoniously prevented by central government move to denude local power and to centralize key decisions over finance and service-delivery administration. Recent legislation and (perhaps more pervasively) *advice notes* from central government departments have had profound constraining impacts on the discretion of local authorities to plan for and deliver transport, primary education and council housing in rural areas. Central government perceptions of the problematic therefore appear to give priority to fiscal restriction, power centralization, and ultimately through recourse to market forces, the maintenance of appropriate conditions for capital accumulations. Although market regulation in favour of particular capital and class fractions is feasible within these priorities, it is unlikely that interventionary regulation on behalf of the deprived classes is compatible with these central state functions.

Analysis of policies and plans for rural people in Britain, therefore, should take full account of the sociopolitical constraints which deter the implementation of many policy options. The primary constraint is that imposed by the overwhelming relationship between the state and civil society. If planning is to be viewed as neutral, apolitical and objective, then the state–society relationship must somehow be explained away in such a way as to permit planning unfettered scope for neutral, apolitical and objective action. If, however, planning is viewed as an integral function of a state whose purpose is not only to preserve the status quo (and thereby preserve contemporary inequalities within that status quo) but also to engineer political economic changes for the further benefit of capital interests, then policies and plans must be fully recognized as being part of these overall functions. If the latter option is accepted, even if only partially, then planning activity can be seen to be subject to very limiting constraints in its scope for action.

Beyond the state–society relationship there are a series of secondary relationships which also serve to constrain the discretion available to planning:

(a) *central–local relationships*, the emphasis here being on an increasing centralization of power in Britain;
(b) *private–public sector relationships*, within which the propensity of Thatcherism is to afford greater freedom to the private sector in response to the underlying power of capital interests;
(c) *interagency relationships* which have displayed increasing complexity as the idea of multifunctional economic, service or welfare delivery agencies has become increasingly abandoned in favour of the 'divide-and-rule' benefits of a multiplicity of small agencies, each with limited power.

An awareness of these primary and secondary constraints is vital for any understanding of rural policies and plans in Britain. In 1983, I outlined four political options for planning for the rural needy (Cloke 1983: 350):

● continue with a *laissez-faire* approach, with conservationism, resource rationalization, free-market conditions and rampant gentrification;
● recognize the need for some additional opportunity provision to tackle rural disadvantage, but only within strict budgetary limitations with the emphasis on promoting self-help;
● bring about a radical resource reallocation in favour of measures directed to the opportunity deficiencies suffered by disadvantaged groups;
● sponsor policies of personal subsidy and wealth redistribution, realiz-

ing that even if a combination of government action, subsidy and self-help were able to provide a satisfactory opportunity base in rural areas for all rural residents some of the underlying causes of rural deprivation will remain.

The Thatcher era has reaffirmed the *laissez-faire* approach, perhaps with trappings of the low-cost self-help strategy, as the dominant policy direction. To change this direction it will take not only a major shift in political will but also a degree of political conflict sufficient to create greater levels of autonomous central state action beyond the bounds of the support of capital interests.

Notes

1 Calculated as incomes up to 139% of supplementary benefit entitlement.
2 Council for Small Industries in Rural Areas.
3 Ministry of Agriculture, Fisheries and Food.

References

Ayton, J. 1980. Settlement policies can bring stabilization. *The Planner* **66**, 98–9.

Beresford, T. 1975. *We plough the fields*. Harmondsworth: Penguin.

Blunden, J. & N. Curry (eds) 1985. *The changing countryside*. London: Croom Helm.

Bradley, T. & P. Lowe (eds) 1984. *Locality and rurality*. Norwich: GeoBooks.

Centre for Agricultural Strategy 1986. *Countryside implications for England and Wales of possible changes in the common agricultural policy*. Reading: CAS.

Champion. A.G. 1981. Population trends in rural Britain. *Population Trends* **26**, 20–3.

Cloke, P.J. 1979. *Key settlements in rural areas*. London: Methuen.

Cloke, P.J. 1983. *An introduction to rural settlement planning*. London: Methuen.

Cloke, P.J. 1985. Counterurbanisation: a rural perspective. *Geography* **70**, 13–23.

Cloke, P.J. (ed.) 1987. *Rural planning: policy into action?* London: Harper & Row.

Cloke, P.J. & J.K. Little 1986. The implementation of rural policies: a survey of county planning authorities. *Town Planning Review* **57**, 265–284.

Cloke, P.J. & J.K. Little 1987a. Rural policies in the Gloucestershire structure plan: I – a study of motives and mechanisms. *Environment and Planning A*.

Cloke, P.J. & J.K. Little 1987b. Rural policies in the Gloucestershire structure plan: II – implementation and the county-district relationship. *Environment and Planning A*.

Cloke, P.J. & J.K. Little 1989. *The rural state?* Oxford: Oxford University Press.

Cloke, P.J. & C.C. Park 1985. *Rural resource management.* London: Croom Helm.

Cloke, P.J. & D.P. Shaw 1983. Rural settlement policy in structure plans. *Town Planning Review* **54**, 338–54.

Derounian, J. 1980. The impact of structure plans on rural communities. *The Planner* **66**, 87.

Development Commission 1984. *Guidelines for joint rural development programmes.* London: HMSO.

Development Commission 1985. *Rural Development Programmes: further guidance from the Commission.* London: HMSO.

Edwards, A. & A. Rogers (eds) 1974. *Agricultural resources.* London: Faber & Faber.

Fothergill, S. & G. Gudgin 1982. *Unequal growth: urban and regional employment change in the U.K.* London: Heinemann.

Fothergill, S. & G. Gudgin 1983. Trends in regional manufacturing employment: the main influences. In *The urban and regional transformation of Britain,* J.B. Goddard & A.G. Champion (eds), 27–50. London: Methuen.

Gilg, A.W. 1984. Politics and the countryside: the British example. In *The Changing Countryside,* G. Clark, J. Groenendijk & F. Thissen (eds), 251–60. Norwich: GeoBooks.

Gilg, A.W. (1985a). *An introduction to rural geography.* London: Edward Arnold.

Gilg, A.W. (1985b). *Countryside Planning Yearbook.* Norwich: GeoBooks.

Gloucestershire County Council and Cotswold District Council 1981. *Rural Action: Brief and Programme.* Gloucestershire County Council, Gloucester.

Goldsmith, M.J. 1986. *New research in central–local relations.* Farnborough: Gower.

Green, R.J. 1971. *Country planning.* Manchester: Manchester University Press.

Healey, P. 1982. Understanding land use planning: the contribution of recent developments in political economy and policy studies. In *Planning Theory: Prospects for the 80's.* P. Healey, G. McDougall & M.J. Thomas (eds), 180–93. Oxford: Pergamon.

Heath, A., R. Jowell & J. Curtice 1985. *How Britain votes.* Oxford: Pergamon.

Keeble, D.E. 1980. Industrial decline, regional policy and the urban–rural manufacturing shift in the United Kingdom. *Environment and Planning A* **12**, 945–62.

Lowe, P., T. Bradley & S. Wright 1986. *Deprivation and welfare in rural areas.* Norwich: GeoBooks.

Massey, D. & R. Meegan 1982. *The anatomy of job loss.* London: Methuen.

McLaughlin, B.P. 1986. The rhetoric and the reality of rural deprivation. *Journal of Rural Studies* **2**, 291–308.

Newby, H. 1979. *Green and pleasant land?* Harmondsworth: Penguin.

Newby, H. 1981. Urbanism and the rural class structure. In *New perspectives in urban change and conflict,* M. Harloe (ed.), 220–43. London: Heinemann.

Pacione, M. 1984. *Rural geography.* London: Harper & Row.

Pettigrew, P. 1987. A bias for action: industrial development in Mid Wales. In

Rural planning: policy into action? P.J. Cloke (ed.), 102–121. London: Harper & Row.

Phillips, D. & A. Williams 1984. *Rural Britain: a social geography*. Oxford: Blackwell.

Pye-Smith, C. & R. North 1984. *Working the land*. London: Temple Smith.

Rees, G. 1984. Rural regions in national and international economies; In *Locality and rurality*. T. Bradley & P. Lowe (eds). Norwich: GeoBooks.

Robert, S. & W.G. Randolph 1983. Beyond decentralisation: the evolution of population distribution in England and Wales, 1961–1981. *Geoforum* **14**, 75–102.

Ryder, J. & H. Silver 1985. *Modern English society*, 3rd edn. London: Methuen.

Smart, G. 1987. Co-ordination of rural policy-making and implementation. In *Rural planning: policy into action?* P.J. Cloke (ed.), 200–212. London: Harper & Row.

Wenger, C. 1980. *Mid Wales: development or deprivation*. Cardiff: University of Wales Press.

Williams, G. 1984. Development agencies and the promotion of rural community development. *Countryside Planning Yearbook* **5**, 62–86.

Woodruffe, B.J. 1976. *Rural policies and plans*. Oxford: Oxford University Press.

Working Party on Rural Settlement Policies 1979. *A Future for the Village*. Bristol: HMSO.

3 The Netherlands

JAN GROENENDIJK

The approach to Schiphol airport in the Netherlands by air affords fascinating views of a meticulously arranged countryside. The development of towns and villages has been neatly contained; the land has been efficiently parcelled by reclamation or land consolidation schemes. The view gives the distinct impression that one is landing in a planned society where individuals relinquish their property rights for purposes of public coordination (Cox 1973). The best imaginable results of physical planning seem to have been achieved in the Netherlands.

But, back on firm ground, the objective of physical planning clearly goes further than merely rearranging that which can be seen. A current definition (Drupsteen 1983) in use by the central government (Commissie Interdepartementale Taakverdeling en Coödinatie Betuursorganisatie, 1971) indicates the goal of physical planning to be 'optimizing mutual adjustment of space and society on behalf of society's well-being'. This implies that policy objectives seek the reallocation of resources and relocation of activities. Indeed, several reports on physical planning explicitly formulate a policy of redressing spatial living conditions on behalf of disadvantaged localized groups. In a mixed economy, some of the consequences of a pure market economy must be ameliorated to reach a more equitable situation.

This chapter attempts to analyse what this means for the rural areas. We have to bear in mind that the perception of policy options is constrained by an acceptance of the status quo (Cloke & Hanrahan 1984). In a more practical sense it is also constrained by the fact that policies are prepared by government departments and their agencies, acting according to the prerogatives of their well defined domain and tradition, and by the fact that consensus on policy within the government is reached by pulling strings and by trade-offs in bureaucratic politics (Allison 1969).

Secondly, we follow the path of decision making from policy formulation at one level of government to its implementation at another. For what is supposed to be merely implementing the decisions according to the specific local context, scores of minor decisions are required, influenced by the local power structure. These decisions might well run counter to earlier policy formulations, and are particularly important when analysing the actions that reach beyond the cosmetic repair and upkeep of a beautiful countryside.

To demystify the planning success story and to arrive at more appropriate expectations, we start by uncovering some traits of Dutch environmental management that give the countryside its orderly appearance. In the first place, ever since the reclamation of the fenlands in the Middle Ages, in the province then known as Holland, polder boards have been established by the co-operating landholders to control the water level. In many respects they have to abide by laws. Polder boards have become institutionalized nationwide as a functional level of local government. The polder board levies taxes on all residents, but the vote is only extended to the landholders. This is one way in which agricultural entrepreneurs influence the environment. Secondly, money from Holland's rich cities was partly invested in the countryside. In particular it was used to reclaim land from the many lakes (partly created by digging peat). These polders are characterized by their geometrical precision in the service of rationalized agriculture. The recently reclaimed Ijsselmeerpolders have been arranged according to detailed landscape plans. Though extensive, their area was surpassed by the 19th and 20th century reclamation of the moors in the East and South. The remaining territory, except for a few nature preserves, has been modernized by land consolidation schemes, which have even been implemented for a second time to keep pace with agrarian modernization.

Finally, the containment of the extension of settlements is explained partly by the timely adoption of town and country planning. However, as nearly all construction sites are at some point publicly owned, the extensions connect smoothly to areas previously built up. Municipalities buy the property, service it and sell it to investors. As a consequence, the contrast between the densely developed villages in the Netherlands and the wide scattering of new houses in Belgium is striking (Groenendijk and Mast 1984). The rapidly expanding cities in the west were the innovators of public property development. In view of the weak structure of the soil, development had to encompass large areas to be efficient. This innovation was subsequently adopted all over the country and eventually turned all municipalities into developers.

Problems of rural areas

The relatively small size of the Netherlands does not preclude regional differences or the influence of distance. To describe social inequality, Engelsdorp Gastelaars *et al.* (1980) classified 'daily activity' regions according to variables of the material well-being of residents on a centre–periphery scale. This produces a pattern of central regions in the three Randstad provinces (Noord-Holland, Zuid-Holland and Utrecht) and (semi-)peripheral regions in the rest of the country. In between are the regions with an intermediate status; these are tied to urban centres outside Randstad such as Breda, Eindhoven, Arnhem, Deventer and Groningen. Regional policy in the postwar era first tried to decrease inequality by advocating industrialization of relatively small regions with high unemployment rates. Some of the industrialization in urbanized villages and small cities scattered throughout the peripheral rural regions dates from this period. Indeed, as Wever (1986: 150) observes, 'until the end of the sixties regional policy was mainly focused on rural agricultural areas'. Later policy included urbanized 'restructuring' areas and was no longer restricted to industrialization but aspired to foster city-regions in the North and South. It is doubtful, however, whether this policy has had a positive effect on the problems of relative deprivation in rural areas.

Combining several variables of material well-being used in the last census (quality of housing, car ownership, etc.), the Sociaal en Cultureel Planbureau (1980) calculated deprivation scores for municipalities and neighbourhoods (Table 3.1). Bearing in mind the small size of rural municipalities, the lowest ranking cannot fail to contain only rural municipalities, although urban deprivation only shows up at the neighbourhood level. Of the 10% lowest-ranking municipalities, Groningen has by far the largest share (partly due to the small size of the municipalities), followed by the peripheral provinces of Friesland and Zeeland. Another part of the country with a large share is the area of the major rivers in the southern parts of Gelderland and Zuid-Holland; these are zones to be kept free of urbanization, according to the policy maintained over the last few decades, although they apparently house many disadvantaged people in remote villages.

In spite of its overall high population density, the Netherlands has relatively open stretches of rural land that lack settlements with urban facilities. Many facilities are not within easy reach for their target categories (housewives, schoolchildren, the elderly). Constraints on the residents' time and resources often prohibit them from undertaking more than the most essential activities (Huigen 1986). This explains why distance travelled in rural areas is only slightly higher than in urban areas (Rijksplanologische Dienst 1984: 18). In the same way, the public expenditure on public transport for the residents of large cities and commuter areas is much greater than that for the population of rural areas (SCP 1984).

Several recent developments in society will only exacerbate the peripheral situation of the rural population. Cuts in employment mean that work may have to be accepted at greater distances from the home. Suitable occupations for parents of young children are scarce in rural areas. Further education or part-time education is difficult, especially when this entails travelling large distances in the evening.

Table 3.1 Municipalities by classes of deprivation scores per province.

Province	1.735–0.951	0.950–0.728	0.724–0.593	0.590–0.427	0.427–0.278	0.278–0.136	0.135––0.031	-0.031––0.270	-0.275––0.628	-0.630––2.714	Total
Groningen	24(47%)	10(17%)	5(10%)	3(6%)	2(4%)	3(6%)	2(4%)	1(2%)	1(2%)	1(2%)	51(100%)
Friesland	9(21%)	6(14%)	8(19%)	4(9%)	7(16%)	4(9%)	2(5%)	2(5%)	1(2%)	-	43(100%)
Drenthe	1(3%)	7(20%)	4(12%)	7(20%)	3(9%)	4(12%)	1(3%)	3(9%)	3(9%)	1(3%)	34(100%)
Overijsel	7(13%)	10(18%)	7(13%)	7(13%)	5(9%)	7(13%)	4(7%)	1(2%)	4(7%)	3(5%)	55(100%)
Gelderland	14(13%)	17(16%)	12(11%)	14(13%)	13(12%)	8(8%)	6(6%)	9(8%)	7(7%)	6(6%)	106(100%)
Utrecht	-	3(6%)	5(11%)	2(4%)	2(4%)	3(6%)	4(9%)	6(13%)	7(15%)	15(32%)	47(100%)
Noord-Holland	1(1%)	1(1%)	3(3%)	4(4%)	12(11%)	11(10%)	18(17%)	17(16%)	23(21%)	23(21%)	107(100%)
Zuid-Holland	12(8%)	10(7%)	9(6%)	10(7%)	10(7%)	11(7%)	14(9%)	15(10%)	32(22%)	26(17%)	149(100%)
Zeeland	8(20%)	4(10%)	4(10%)	5(13%)	5(13%)	3(8%)	3(8%)	5(13%)	2(5%)	-	39(100%)
Noord-Brabant	10(7%)	9(7%)	20(15%)	16(12%)	14(10%)	16(12%)	15(11%)	6(4%)	10(7%)	10(7%)	136(100%)
Limburg	1(1%)	10(9%)	11(10%)	10(9%)	13(12%)	20(19%)	17(16%)	14(13%)	8(8%)	2(2%)	106(100%)
total municipalities	87(10%)	87(10%)	88(10%)	87(10%)	87(10%)	88(10%)	87(10%)	87(10%)	88(10%)	87(10%)	873(100%)

Note
Calculated from 'List of municipalities by degree of social deprivation' (SCP 1980: 81).
Negative scores indicate relative absence of deprivation.

The redirection of the welfare state which is currently underway does not present a rosy prospect either. At the time the Rural Areas Report (Ministerie van Volkshuisvesting en Ruimtelijke Ordening, 1977) was prepared, there were still signs that central government would engage in measures of compensation for the problems of scale that complicate the provision of services in rural areas. The Spatial Perspectives Memorandum (*Ruimtelijke Perspectieven*), an interim report pending release of the Fourth Report on Physical Planning, (Rijksplanologische Dienst 1986) relies on the strong parts of the country to meet international competition. As stated in a memorandum (May 1986) from the National Union of Small Settlements, general measures of austerity towards social services will have their worst effects in rural areas. By incorporating the preschool and kindergarten into the primary school more than 50 'last schools of the village' have already been closed and 50 to 60 more will follow within the next two years, preventing young households from settling in those villages. Rationalization of public transport, postal services and libraries is mentioned as an other example of economy measures applied without recognition of the specific situation of rural areas.

To understand what policies may effectively be implemented, we have to determine what agencies are designated to carry them out; then we can analyse each particular formulation of policy in terms of its inherent priorities. In addition, its success or failure is better understood when the political culture of rural areas is analysed. This chapter now turns to these issues, ending with an analysis of the implementation of policy measures in the local environment in two case studies.

Agencies and tiers of government: their interrelations

The position of physical planning among the tasks of government has been a matter of much concern, especially to the Ministry of Housing, Planning and Milieu (HPM). A model of the relations is shown in Figure 3.1.

This model is easily recognized in published policies. Physical planning as a facet of planning was established in the Third Report, with sections on urbanization and rural areas (Ministerie van Volkshuisvesting en Ruimtelijke Ordening 1976, 1977). In joint responsibility, the sector departments and HPM have generated a stream of structure schemes for water supply, traffic and transport, defence areas, open air recreation, nature and landscape preservation, land use, etc. (Ministerie van Landbouw 1981; Ministerie van Cultur, Recreatie en Maatschappelijk werk 1981a, 1981b). Figure 3.2 shows which agencies perform planning tasks within distinct tiers of government: the Physical Planning Agency (Rijksplanologische Dienst) within the HPM, a committee representing other departments and an advisory council as a forum for experts and interest groups.

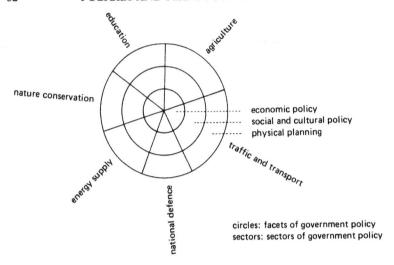

Figure 3.1 Sector and facet planning.
Source: Information and Documentation Centre for the geography of the Netherlands.

The other tiers of government are responsible for their own policy, provided they do not conflict with plans of higher authorities or are overruled by a formal directive. The threat of intervention will usually evoke a compromise. On the provincial level, the Provincial Physical Planning Committee has several representatives of sector departments among its members. The HPM is represented by one representative for housing and one for planning. By making regional plans (*streekplannen*) and by supervising municipal planning, provinces co-ordinate the planning that takes place at the three levels. This co-ordinating task is generally delegated to the Provincial Planning Department (PPD). Municipalities have to draw up development plans for their entire territory (and have the option of additional plans for built-up areas). These plans have legal implications for the public in that they state exactly what use is to be made of land and buildings. This binding character entails a thorough and lengthy procedure to prevent undue damage to those (landed!) interests on which it may intrude. In this way a plan cannot depart too far from the existing situation unless it benefits the owner.

In principle, this system connects departmental measures on the one hand with policies of local governments on the other. However, by virtue of their control over budgets, sector departments can decide whether or not to use the opportunities available within the constraints of physical plans. In the central–local relationship, by their 'local presence'

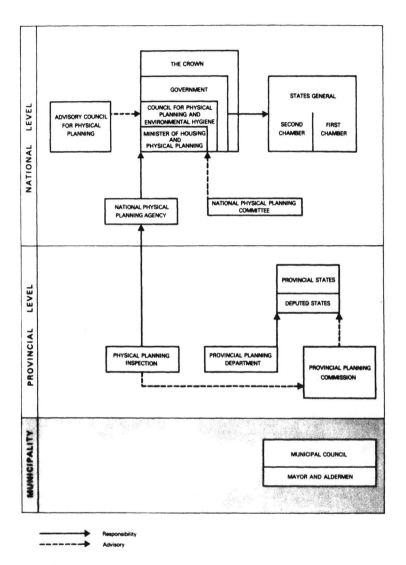

Physical planning agencies.

Figure 3.2 Physical planning agencies.
Source: Information and Documentation Centre for the geography of the Netherlands.

municipalities take the lead in terms of information, which places them in a position to co-ordinate sector departments 'from below' (Toonen 1981).

The sectoral arrangement of central government implies that there are no agencies that occupy themselves specifically with rural areas. Only agriculture now has a Directorate of Rural Areas; it includes agencies that were committed to open space in the 1982 Cabinet formation, but these do not include social issues. Indeed, the idea of the decentralized unitary state implies that general policies are applied locally by municipalities, tailored to the particular requisites of the environment. However, the small rural municipalities are understaffed; local integration of the many specialized directives issued by the funding sector departments is therefore not optimal. On the other hand, somewhat larger rural municipalities are likely to give priority to the development of their central village, aspiring to a regional function (de Boer & Groenendijk 1986). In this way, rural areas lack a governmental institution appropriate to their political needs.

Policy content

The outright priority given to urbanization – as formulated in regional policy – means that hardly any thought has been given to the future of rural life, except for the concentrated production functions. At one time provinces started to make regional plans for rural areas (Fig. 3.3), but the ideas for rural areas expressed by the highly rural provinces were not elaborated beyond the state of draft reports (PPD Groningen 1962, PPD Friesland 1966). The idea of modernizing the settlement pattern by concentrating new development (industry, housing, services) in a few larger villages figures prominently in these reports. Elsewhere as well, this idea determined the provincial assessment of municipal plans. In fact it became the generally accepted standpoint in most municipalities, or rather in the reports and plans of their consultants (who did not always convince representatives of the small settlements in the municipal councils). The Second Report on Physical Planning (Ministerie van Volkshuisvesting en Ruimtelijke Ordening 1966) concentrated on rural phenomena that are complementary to urban development. Aside from this refined urban modelling, the report only referred to the concentration strategies for rural areas that were generally applied by rural municipalities. The rôle of rural living conditions in forging the fate of small settlements appeared on the political agenda of the Provinces only during the 1970s. In the first part of the Third Report on Physical Planning (the Orientation Report) (Ministerie van Volkshuisvesting en Ruimtelijke Ordening 1974), Groot's (1972) research on village life, which pointed out the importance of the social climate of small villages, was used to suggest the formulation of policies

Provincial plans in power 1965-1980

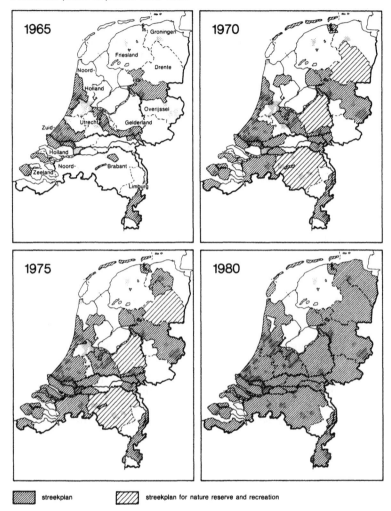

Figure 3.3 Phases in the progress of regional planning: regions for which a *streekplan* has been determined. (Drawn by J. ter Haar.)

which would do justice to the dismay that was expressed at the time in regard to restrictions placed on small settlements. It was suggested that the viability of rural settlements should not deteriorate any further. A way to prevent deterioration was to provide housing for 'the local needs' only, which satisfied those provinces that were anxious about suburbanization. However, even a highly rural province like Friesland long remained firmly entrenched in its policy of 'modernization of settlements pattern' (de Bakker 1984).

The most positive policy ever formulated is to be found in the RAR as part of the Third Report (Ministerie van Volkshuisvesting en Ruimtelijke Ordening 1977). The RAR no longer claims that physical planning is able to bring the solution all on its own and goes on to suggest what sectoral measures should be taken. Most of them are only policy suggestions at the level of sector departments (pooling or mobilizing of dwellings, lowering of thresholds). In this respect it should be noted that the housing sector department of HPM did not provide special requirements for building small amounts of public housing (PH). The spatial pattern of areal functions is analysed along principles of separation and integration. Four landscape zones were distinguished varying from (A) areas with agriculture as the main function to areas with nature as the main function (D) (Fig. 3.4). From the Structure Sketch for Urbanization, the open zones between city regions had to be maintained as such by restricting in-migration to zero or even fostering a negative migration balance. Apart from other restricted zones (zone D and regions that had been growing excessively), provinces and municipalities had to share the responsibility for determining the population distribution.

The impact of the RAR on the sectoral measures and on the plans of local authorities is disappointing; positive effects on housing, services and well-being are hard to find. Indeed, the restrictions on housing have been worked out in regional plans. But in the reigonal plan for the Veluwe (Provincie Gelderland 1985), for example in that part of the region lying out-side the restricted zones, the province gives the municipalities free rein in the distribution of housing. There are no specific co-operative strategies by which municipalities can acquire services and facilities for their rural areas. But, of course, central government had hardly anything to offer. A revised Structure Sketch for Urbanization (Ministerie van Volkshuisvesting en Ruimtelijke Ordening 1983) redirected the policy towards existing city regions (de Smidt 1984). In Spatial Perspectives (SPR, Rijksplanologische Dienst 1986) it is announced that the Fourth Report will comprise only one Sketch for urban and rural areas. This integration is understandable; but though production activities predominate in this report, its thrust is the position of the Netherlands as a major agricultural exporter and the relationship of agriculture to landscape and milieu (the domain of the agency newly acquired by the department).

Impact of policy: problems of implementation

In order to understand policy implementation for the rural areas it is first important to consider the administrative environment in which policies are

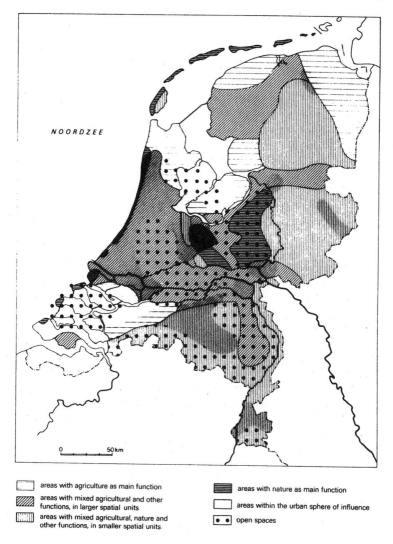

NOORDZEE

	areas with agriculture as main function
	areas with mixed agricultural and other functions, in larger spatial units
	areas with mixed agricultural, nature and other functions, in smaller spatial units
	areas with nature as main function
	areas within the urban sphere of influence
• •	open spaces

0   50 km

Figure 3.4   Summary of the Structure Sketch for the rural areas. (Drawn by J. ter Haar.)

to be implemented. Several types of relationship may be discerned in the administrative structure of the Netherlands. On the one hand, there are ministries with their own implementing agencies and subagencies throughout the country: the Ministry of Traffic and Public Works; the Ministry of Agriculture, with its staff for changing land use. On the other hand, HPM has offices in every province, but the ownership of public housing (PH) is left to the 'quasi non-governmental' housing associations (HA). The implementation of higher authorities' plans tends to be characterized by compromise. This is very much the case in the relationship between the province and the municipality. The province has the task of supervision but with its low budget, it cannot directly influence the resource management strategy of the municipality. For example the Province of Friesland changed its policy to spread the population (Provincie Friesland 1981), only to discover that the municipalities did not comply with the plan, mainly because of long-term investment interests (PPD Friesland 1984). Secondly, it is important to consider the societal environment in which policy is implemented. In a market economy this forms a constraint both on the policy options and on the results. In the province of North Holland, for example, concentration of resources in larger villages was mainly realized by the production of houses, but not by investment in commercial services or by migration preferences of the regional population (PPD Noord-Holland 1984). In the regional plan for the Veluwe, it is clear that further restrictions on housebuilding will only exacerbate housing need, inducing people to make clandestine use of summer homes, caravans, etc. Areal policies aiming at integrating agriculture and nature preserves have failed to date (Rijksplanologische Dienst 1986). Market forces and EEC policies that decrease prices induce the agrarian population to reduce their input costs by intensifying their operations in some way. According to Benvenuti (1975), integration of the farmer into agribusiness complexes and a wider 'technical administrative task environment' constrains his options, making him more the executor of predefined tasks than the entrepreneur which he essentially is. According to Table 3.2, the smaller farms in zones C and D are particularly pressured to intensify production. This in effect pre-empts the integration foreseen for these zones.

Political culture

The Netherlands is no exception to that general feature of Western democracies whereby conservative political parties are overrepresented in rural areas, although Table 3.3 shows that the share of left-wing parties in the North is larger than elsewhere. Of crucial importance, however, is to have a seat in the Court of Burgomaster and Aldermen. This Court forms the executive, bears the responsibility for the portfolios that make up municipal tasks and in this way determines municipal policy. Denters (1981) demonstrated a positive correlation between representation of aldermen from left-wing parties and the share of public housing in the housebuilding programme. According to Table 3.3 this relation may be

Table 3.2 Characteristics of agricultural production per zone of the Rural Areas Report 1982.

	Netherlands	A	B	C	D
number of farms	139.654	25.868	47.741	56.776	9.269
mean area per farm (ha)	14.4	23.0	12.4	13.0	8.7
mean production per farm (SPE[a])	159	177	177	143	114
dairy cattle per ha	2.8	2.4	2.6	2.9	2.8
percentage yearly increase of dairy cattle per ha (1981, 1982)	2.4	1.5	2.3	2.5	4.0
production of farmyard manure (in 1000 kg/ha)	25.9	8.2	28.8	36.4	39.2
production of nitrogen (kg/ha)	141.2	41.0	153.1	204.2	212.6
production of phosphate (kg/ha)	88.2	23.1	92.8	131.0	140.6

[a]) Standard production entities. In 1984 the frequency distribution in the Netherlands over classes of SPE in quartiles was as follows (farming as main occupation only): 1st quartile up to 90 SPE, 2nd 90–150, 3rd 150–250, 4th 250–500 (with the exception of 4%>500 SPE).

Source: RPD 1984.

extended to rural municipalities where representation of leftist aldermen is mainly found in the North.

Several interests may be served by the rural municipality as far and as long as its autonomy is retained. These are primarily agrarian interests. In contrast to the position of the large landholder described by Newby *et al.* (1978), in the Dutch local power structure, it is the ordinary farmer (see Table 3.2), in his entrepreneurial rôle (with help from his organization and official research institutes) who has a tremendous impact on the local implementation of policy concerning agrarian interests. As it happens, both the farmer and the political arena of the rural municipality in which he operates are, on average, rather small.

Local policy-makers often strive to preserve the character of the village or the municipality as an entity (Instituut Contact Randgemeenten 1969, van Ruller 1972). From examples in peri-urban situations, it is clear that this goal does not exclude the possibility of the gentrification of existing areas or construction of residential quarters for urbanites. Gentrification furthers the interests of the resident population by inducing new investment, generating many more positive effects than PH would bring.

By amalgamating municipal territories, these interests might be at stake.

Table 3.3 Share of votes and share of aldermen for left-wing parties in municipalities according to degree of urbanization per province in 1986.

| | Percentage of votes in type of municipality | | | Percentage of aldermen | | Public housing as percentage of total housing units built 1973–84 in rural municipalities (A) |
	Rural (A)	Urban/rural (B)	Urban (C)	(A)	(A+B+C)	
Groningen	43	47	55	44	56	43
Friesland	36	38	48	50	50	40
Drenthe	41	37	48	45	47	28
Overijssel	20	27	42	29	33	30
Gelderland	34	27	40	26	33	30
Utrecht	19	24	41	5	21	36
Noord-Holland	28	30	50	22	35	31
Zuid-Holland	17	30	45	13	36	42
Zeeland	25	23	38	32	35	35
Noord-Brabant	4	17	34	3	18	25
Limburg	13	14	33	7	15	23
Nederland	26	26	43	26	32	35

Note: Degree of urbanization: criteria of 1960 applied to census of 1971.
Source: Centraal Bureau voor de Statistiek 1984 (Statistiek der Verkiezingen 1982). Den Haag: Staatsuitgeverij, 1984.

Depending on the goals of such an operation, urban–rural opposition could become an important issue. But by reducing the reorganization of local government to the integration of small municipalities, the status quo seekers have already made their point. Reorganization of local government on a regional basis, with a number of key responsibilities being transferred to the regional level – as has been the intention for a long time – would have widened the local political arena to include the nodal city.

One of the main objectives of territorial structuring is to improve the hierarchy of services by assigning rural areas to nodal villages or towns. In an evaluation of a decade of municipal reorganization, Van Dijk (1979) shows that most proposals have been amended by Parliament to allow for non-nodal integration. In this way, large 'agrarian' municipalities have been formed. In conclusion, not enough interest has been shown in the creation of regional decision-making platforms for regional development to compensate for opposition to reorganization. It is not clear what has been won for the population of remote villages by those rural areas that managed to remain free from urban influence but did not organize in such a way that rural disadvantages could be effectively remedied.

Local implementation

We now turn to two much-disputed issues that have had great impact on the life of the rural population: housing and the preservation of nature and landscape. The municipality has the well established responsibility in the provision of (new) housing that goes far beyond the physical planning task of making development plans and carryng out building-code inspections. The municipality distributes housing in those pressured areas where the Housing Act (WRW) is still in effect. It plays a crucial part in the building of houses by developing the construction sites. Only in one respect has the municipality had to retreat. Since 1969 it may no longer commission PH projects when there is a local housing association (HA) that is prepared to perform the task. Although many municipalities reacted by creating a HA of their own, in many situations the HA operates as an additional actor in the decision-making process. The municipality obtains the financing and designates sites within its territory. This links up with its function as developer; indirectly, this promotes a regional service function for the central village, a status to which many municipalities aspire. Clearly, the higher the population threshold of this central village, the more convincing the claim to services will sound to the ears of higher authorities and investors.

By restricting the provision of housing to the satisfaction of the local needs, the function of rural municipalities for the urban housing market was curtailed. The definition of the concept of 'local need' is, of course, crucial to the implementation of this restriction. To this end the province uses certain criteria (economic or social ties) in its regional plan to calculate the degree of growth allowed in the housing stock or the population. To make municipal development plans comply with provincial policy, only 'global' plans are passed by the province. Only when this global plan is elaborated (art. 11 WRO) under provincial supervision, may planning permission be issued. Elaboration is allowed by the province upon demonstration of a specific housing need. The province of Utrecht, for example, checks this by inspection of 'house-hunter' lists, and in doing so oversteps the domain of physical planning (Kroes & Vulperhorst 1982) as the province may check numbers and disposition of housing projects but not their distribution. Here we cross the line to the responsibilities of the sectors, where the Housing Act charges municipalities with the distribution of housing (van Weesep 1982). The enforcement of this Act is now restricted by liberalization of all but the pressured areas (west and centre of the country) where municipalities ask for reinstatement of the Act. But even then, certain categories of housing (higher-priced) and of population (elderly people that no longer have 'economic ties') are exempt from this regulation. Particularly in some highly valued rural areas, gentrification may take place by the influx of affluent elderly residents, squeezing out young local households. Scarcity of housing, caused by restrictions in the sphere of physical planning, does not permit a fair distribution, nor can the 'local needs' be satisfied; these vowed intentions of planning are frustrated by the lack of appropriate legal powers. The development of building sites for new housing places the municipality in a position to restrict sale to parties that comply with criteria such as economic or social ties. Implementation of the 'local needs' housing policy implies that the rate of addition to the housing stock and the type of houses built should be in accordance with the needs of the local population. For small villages this means that only very small numbers of housing units are to be produced, and a considerable share of these will be PH. The inability to satisfy these requirements within centrally-imposed budgetary constraints highlights the balance of power between central and local government (Vereniging van Nederlandse Gemeenten 1979, Landelijke Stuurgroep BRW 1984).

A number of case studies demonstrate that local political priorities are crucial in this respect (Groenendijk 1986). Only in a minority of municipalities is the distribution of housing over settlements a contested issue. There it is engrained in the political culture and is a self-evident fact in bureaucratic procedures. PH construction projects encompass several sites and small numbers of housing units. On a more general basis, local builders reckon that very small projects develop occasionally in a long-term relationship with municipality and HA. Small building sites are bought when the opportunity arises: the continuity in housing construction instils the confidence that these sites will be needed.

In the majority of municipalities, the administrative obstacles to building in several settlements are circumvented in a strategy of non-decision. When pressure builds up in the municipal board, it is argued that only larger projects which cover investment in the site are feasible. In addition, the HA, responsible for the exploitation of PH, tends to practise risk-avoiding strategies by investing in larger villages, making maintenance less costly as well.

LOCAL IMPLEMENTATION OF POLICY FOR NATURE AND LANDSCAPE RESERVATION

The RAR observed that the countryside is no longer only economically valued for its agricultural production but for the qualities of its landscape as well. Just prior to this observation, a report had been published on the relation between agriculture and nature and landscape preservation (Ministerie van Landbouw 1977) introducing the idea of management contracts with farmers. The Land-use Act (Landinrichtingswet 1985) broadened the scope of land consolidation to include other aims such as landscape preservation. Because this Act focuses on the facet of planned intervention – at least nominally – the province with its regional plan has a central rôle in decision making. It has to put restrictions on agriculture in line with central government's ruling in the structure schemes. These measures can only be enforced by adoption in a Development Plan for areas Outside Settlements (DPOS). The option of making a DPOS has not been popular among municipalities. It does not bring any 'development', only restrictions, and it does not appeal to farmers who are firmly entrenched in local politics. Existing plans made under the old physical planning regulations (*Woningwet*) were less explicit and offered possibilities for issuing planning permission outside settlements in case agrarian landowners would have to be 'bought off'. It is only when old plans no longer allow proper management by the municipality that new plans have been made (Bouwman-Geeraedts & Jansen 1981).

In case studies concerning the area south of Groningen, van der Moolen (1985) reports that municipalities take the side of the farmers. They fight with success the designation of large areas as Relation Report Areas by the province (the instruction of central government). Only half the area originally designated is to be preserved and land consolidation will modernize the rest.

Nearby in Drenthe, the new Land-use Act opened possibilities for agricultural modernization on a partial basis, without demolition of the important non-agricultural values of the environment. In 1984 a land-use committee was inaugurated with Roden's burgomaster as chairman. Meanwhile, a first draft of the regional plan for North Drenthe was published to guide the restructuring and land consolidation plan. More than 50% of the agrarian territory of the municipalities was classified as highly valuable with respect to nature and landscape, which restricted the agricultural use. The reaction of the municipality of Roden was that it already had a structure plan and that the province could not make restrictions on agriculture more severe. In the provincial political arena farmers' interests easily come to the forefront. They are well represented in *provinciale staten* (County councils), except in the left-wing parties. In Drenthe's Council it was decided to greatly reduce the area where agriculture is to be secondary to nature and landscape. (Streekplan Noord-Drenthe, 1986). Instead of directing the Commission's work, this definite version of the regional plan leaves decisions on many points up to the land-use committee.

Conclusions

Provisions for the well-being of the rural population differ significantly from those in urban situations because they require adaptation to low population thresholds. In an era of rising demands the solution to this problem in the Netherlands was sought in physical planning. Spatial co-ordination was to tackle scale problems. In this way, the low incomes prevailing in rural areas were neglected as part of the problem. During the 1970s, poor results showed that the services to be provided in a multitude of sectors were not easily co-ordinated. Professionalization, intended to improve services, led to scale enlargement rather than the reverse. Solutions had to come from the sector departments, allowing municipalities to spend budgets according to a low-key approach in small settlements that were out of reach until then.

In consequence, the RAR took a more modest position in physical planning. For positive steps it referred to scores of sectoral measures to be taken. As has been proved more generally in the Netherlands, however, more success in physical planning has been made in preventing undesired developments than in promoting development where it is wanted. Central government agencies did not initiate scale-decreasing adaptations. Their main professional achievements clearly had to be made elsewhere. Severe cuts in spending precluded further interest in this direction.

Indeed, in matters of 'social consumption' it is local government that is supposed to take the lead. According to Saunders (1981: 43), local government is most susceptible to popular pressure. It wrestles budgets from central government for specific local needs. This is not what we generally see in rural municipalities, where centre-right political inclinations predominate and 'popular pressure' is not characteristic for the local political culture, especially in respect to the fate of the population in small settlements. Furthermore, small rural municipalities are no match for the professional-sector agencies. Somewhat larger municipalities tend to compete with each other for regional functions in their largest settlement rather than occupy themselves with problems of rurality on the periphery of their territory. Burgomasters and aldermen usually manage to keep the latter from the agenda. Support of the agrarian population, which is very well represented on the municipal board, is crucial to this non-decision strategy. Farmers are not dependent on housing provisions and usually secure planning permission for themselves. For them the improvement of conditions for agrarian production is paramount, certainly it is more important than service provision. In response to their influence, the interorganizational compromise reached at national level is resettled locally. This is in striking contrast to the poor results of pressure in the issues of housing and service provision. In contrast to Buchanan's (1982) findings that in the English county of Suffolk there is a planning élite that keeps issues off the political agenda, in the Netherlands the authorities at the county level not only failed to do so but even had to retreat on agricultural issues. This confirms Buller & Lowe's (1982) critique of Newby et al. (1978): there is a clear conflict between agricultural production and nature preservation. In the fragmented layer of municipalities in the Netherlands it is the farmer, or rather the agribusiness complex, that wins. Highly predominant interests in one sector of production induce 'rural managers' in the local political arena to stand behind the agrarian population rather than occupy themselves with the awkward problems of small-scale service provision, and bio-industry occasionally precludes further building of houses. In this respect the 'corporate level' may have priority locally as well (Johnston 1985, Cloke 1986). Evidence from the Netherlands suggests that municipal activity is not merely to be seen as the implementation of planning decisions made elsewhwere. In this respect the differentiation made by Faludi & ten Heuvelhof (1985) between decisions 'in principal' (laid down in plans to improve 'operational decisions' but open to redirection) and 'operational decisions' (which take immediate effect in the environment) seems more fruitful than the policy–implementation divide. Both types of decisions are made by central and local government, and both are politically relevant.

References

Allison, G.T. 1969. Conceptual models and the Cuban missile crisis. The American Political Science Review **LXIII**, 689–718.

Bakker, D.H. de 1984. *Evaluatie van het centrumdorpenbeleid in de provincie Friesland*. Utrecht: Geografisch Instituut.

Benvenuti, B. 1975. General systems theory and entrepreneureal autonomy in farming: towards a new feudalism or towards democratic planning. *Sociologia Ruralis* **XV**, 46–64.

Boer, T. de & J.G. Groenendijk 1986. Local government and small villages in peripheral rural areas: a survey of municipal policy-making. In *Rural development issues in industrialized countries*, G. Enyedi & J. Veldman (eds), Budapest: Centre for Regional Studies, 147–160.

Bouwman-Geeraedts, M.B.P. & A.J. Jansen 1981. *Het buitengebied landelijk bezien*. Leiden: Rijksuniversiteit.

Buchanan, S. 1982. Power and planning in rural areas: preparation of the Suffolk country structure plan. In *Power, planning and people in rural East Anglia*, M.J. Moseley (ed), Norwich: University of East Anglia, 1–20.

Buller, H. & P. Lowe 1982. Politics and class in rural preservation: a study of the Suffolk Preservation Society. In *Power, planning and people in rural East Anglia*, M.J. Moseley (ed.), Norwich: University of East Anglia, 21–41.

Centraal Bureau voor de Statistiek 1984. *Statistiek der Verkiezingen 1982*, Den Haag: Staatsuitgeverij.

Cloke, P.J. & P. Hanrahan 1984. Policy and implementation in rural planning. *Geoforum* **15**, 261–9.

Cloke, P.J. 1986. Implementation, intergovernmental relations and rural studies: a review. *Journal of Rural Studies* **2**, 245–53.

Cox, K.R. 1973. *Conflict, power and politics in the city*. New York: McGraw-Hill.

Denters, S.A.M. 1981. *Volkshuisvestingsbeleid en bestuurlijke organisatie van gemeenten*. University Twente, Technische Hogeschool Twente Enschede.

Drupsteen, Th.G. 1983. Ruimtelijk bestuursrecht (Volume 1). Brussel, Alphen aan den Rijn: Samsom.

Dijk, S. van 1979. *Kernenhiërarchie, motief voor gemeentelijke herindeling?* Amsterdam: Vrije Universiteit.

Engelsdorp Gastelaars, R.E. van, W.J.M. Ostendorf & S. de Vos 1980. *Typologieën van Nederlandse gemeenten naar stedelijkheidsgraad*. The Hague: Central Bureau voor de Statistiek.

Faludi, A. & E.F. ten Heuvelhof 1985. Lokale planning. In *Lokaal Bestuur in Nederland*, W. Derksen & A.F.A. Korsten (eds), Brussel, Alphen aan den Rijn: Samsom/H.D. Tjeenk Willink.

Groenendijk, J.G. 1986. *Implementing rural planning*. Paper presented at the 2nd British–Dutch symposium on Rural Geography, Amsterdam.

Groenendijk, J.G. & P. Mast 1984. Woningbouw en nederzettingsuitleg aan weerszijden van de Nederlands-Belgische grens. *Geografisch Tijdschrift* **15**, 179–89.

Groot, J.P. 1972. *Kleine plattelandskernen in de Nederlandse samenleving*. Wageningen: Agricultural University.

Huigen, P.P.P. 1986. *Binnen of buiten bereik?* Nederlandse Geografische Studies 7, Koninklijk Nederlands Aardrijkskundig Genootschap, Utrecht: Geografisch Instituut.

Instituut Contact Randgemeenten 1969. *Bestuur in agglomeraties*.

Johnston, R.J. 1985. Local government and the state. In *Progress in political geography*, M. Pacione (ed.), London: Croom Helm, 152–176.

Kroes, J.H. & L. Vulperhorst 1982. *Bouwen voor de eigen woningbehoefte in de provincie in de provincie Utrecht*. Delft: RIW-Instituut voor Volkshuisvesting-sonderzoek.

Landelijke Stuurgroep BRW experiment, leefbaarheid in nieuwe kernen 1984. *Inzicht Uitzicht*. Rijswijk: Ministerie Van Welzijn, Volksgezondheid en Cultuur.

Ministerie van Landbouw 1977. *Nota over de Relatie tussen de Landbouw en Natuur-en Landschapsbehoud* (Report on the relation between agriculture and the preservation of nature and landscape). Den Haag: Staatsuitgeverij.

Ministerie van Landbouw 1981. *Structuurschema voor de Landinrichting* (Structure scheme for Land arrangement). Den Haag: Staatsuitgeverij.

Ministerie van Volkschuisvesting en Ruimtelijke Ordening 1966. *Tweede Nota over de Ruimtelijke Ordening* (Second Report on Physical Planning). Den Haag: Staatsuitgeverij.

Ministerie van Volkshuisvesting en Ruimtelijke Ordening 1974. *Orienteringsnota* (Orientation Report). Den Haag: Staatsuitgeverij.

Ministerie van Volkshuisvesting en Ruimtelijke Ordening 1976. *Verstedelijkingsnota* (Urbanization Report). Den Haag: Staatsuitgeverij.

Ministerie van Volkshuisvesting en Ruimtelijke Ordening 1977. *Nota Landelijke Gebieden* (Rural Areas Report (RAR)). Den Haag: Staatsuitgeverij.

Ministerie van Volkshuisvesting, Ruimtelijke Ordening en Milieu 1983. *Structuurschets voor de Stedelijke Gebieden* (Structure Sketch for Urban Areas). Den Haag: Staatsuitgeverij.

Ministerie voor Cultuur, Recreatie en Maatschappelijk werk 1981. *Structuurschema voor Natuur-en Landschapsbehoud* (Structure Scheme for Nature and Landscape Preservation). Den Haag: Staatsuitgeverij.

Moolen, B. van der 1985. *Het besluitvormingsproces rond de reservaatsvorming in de agrarische gebieden van Haren, Peize-Vries en Roden-Norg*. Groningen: Geografisch Instituut.

Newby, H., C. Bell, D. Rose & P. Saunders 1978. *Property, paternalism and power: a study of farmers in East Anglia*. London: Hutchinson.

PPD Noord-Holland 1984. *Voorzieningenniveaus in West-Friesland*. Studierapport 29, PPD Noord-Holland, Haarlem.

PPD Groningen 1962. *Een onderzoek naar de spreiding van de bevolking en de voorzieningen in Noord-Groningen*. Groningen: PPD Groningen.

PPD Friesland 1966. *De ruimtelijke ontwikkeling van het Friese platteland*. Leeuwarden: PPD Friesland.

PPD Friesland 1984. *Bouwen in Bokwerd*. Leeuwarden: PPD Friesland.

Provincie Gelderland 1985. *Voorontwerp herziening streekplan Veluwe*.

Provincie Friesland 1981. *Streekplan Friesland*.

Provincie Drenthe 1986. *Streekplan Noord-Drenthe*.

Rijksplanologische Dienst 1984. *Voortgangsrapport, nota landelijke gebieden*. Den Haag: Staatsuitgeverij.

Rijksplanologische Dienst 1986. *Nota ruimtelijke perspectieven*, Den Haag: Staatsuitgeverij.

Ruller, H. van 1972. *Agglomeratieproblematiek in Nederland*. Alphen aan den Rijn: Samsom.

Saunders, P. 1981. Community power, urban managerialism and the local state. In *New perspectives in urban change and conflict*. M. Harloe (ed.), London: Heinemann Educational, 27–49.

Smidt, M. de 1984. De stad centraal en afscheid van spreidingsbeleid. *Geografisch Tijdschrift* **18**, 59–65.

SCP (Sociaal en Cultureel Planbureau) 1980. *Sociale achterstand in wijken en gemeenten*. SCP cahier 14, Rijswijk.

SCP 1984. *Sociaal en cultureel rapport*. Rijswijk: SCP.

Toonen, Th.A.J. 1981. Gemeentelijke invloed in een vervlochten bestuur. *Beleid en Maatschappij* **11**, 334–42.

Vereninging van Nederlandse Gemeenten 1979. *Wie het kleine niet eert . . .* Den Haag

Weesep, J. van 1982. *Production and allocation of housing, the case of the Netherlands*. Geografische en Planologische Notities 11, Vrije Universiteit, Amsterdam.

Wever, E. 1986. Regional policy in the Netherlands. *Tijdschrift voor Economische en Sociale Geografie* **77**, 149–53.

4 Scandinavia

PETER SJØHOLT

The Scandinavian rural legacy

Scandinavia is the oldest geographical concept for the northernmost part of Western Europe and is, as such, somewhat imprecise in scope. In its maximum extent Scandinavia comprises all the four Nordic countries – Denmark, Finland, Norway and Sweden. Often, however, the designation is confined to the Scandinavian peninsula, covering Norway and Sweden.

Greater Scandinavia is an area of great physical heterogeneity, Denmark belonging to the Mesozoic Middle European Plain, Sweden and Finland mostly to a low lying area of archaic Precambrian rock and Norway to a mixture of the latter and more rugged Palaeozoic bedrock. This means generally a growing scarcity of land as one proceeds northwestwards and an abundance of fertile soil in the south and south-east. As the climatic conditions generally become harsher from south to north, Denmark and Sweden thus share the best conditions for agriculture and Norway the worst. To a certain extent these resource differentials are compensated for by other resources – fish in Norway and forests in Finland. In a broad context, however, Sweden is the country with the most balanced access to both agricultural and basic industrial resources.

The Scandinavian world is generally culturally and politically homogenous. Except for Finland, languages are similar, and all countries have come to share fairly common sociopolitical values.

In spite of the similarities, however, a treatment of rural policies and plans for the whole area is not unproblematic. There are obvious differences both in the structural basis and political solutions. The very concept of 'rural' is, furthermore, difficult to define across the frontiers when trying to give it a precise meaning.

The author is prone to give it a rather broad definition, referring by the term 'rural' to local communities which, up to the post World War II period, were directly dependent either on areal resources (on-shore and off-shore) or localized natural resources (ores and minerals) for their survival and consolidation.

Broadly speaking, there exist great similarities between the countries in the development of the cultural landscape and the rural local communities.

Overall population density is, except for Denmark, very low. All the countries are made up of old rural core areas which came to serve as the foci and cradles of later urbanization. Surrounding these are the more or less peripheral areas, some of which, except for Denmark, may be classified as extreme 'peripheries'. The settlement of some of the peripheries dates far back in history. But in all countries wide areas in the periphery were colonized fairly recently. Among the latter regions can be mentioned the heather areas in Jutland, earlier virgin forests in northern Finland and Sweden, some south-west Norwegian morainic and heather areas and large tracts in North Norway. All these were settled mainly in the 19th century and some colonization has been going on well into the 1900s, particularly in Finland and Norway.

But there are clear differences in both the natural and cultural rural heritage. Partly these are topographic in nature, distinguishing Norway with its broken relief and formerly isolated communities from most other Nordic regions. Simultaneously, the earlier, relatively concentrated landed property in Denmark and South Sweden contrasts markedly with the more fragmented and egalitarian structure of tenure in Finland and, particularly, Norway. Although the structure had changed towards family farms both in Denmark and Sweden by the 20th century and the areas of colonization in all countries had been characterized by small-scale tenure from the outset, there are remnants of the old organization pattern, manifested in a larger-scale structure in Denmark and southern Sweden than in the rest of the Scandinavian regions.

Up to and, in some regions, well into post-World War II times, the industrialization of the countries and the urban transition depended on direct utilization of natural resources *in situ* by the application of medium to advanced technology. This was exemplified by meat-processing and cement industries in Denmark, wood processing in all the other countries, metal processing in Sweden and fish processing and primary metal manufacturing in Norway. This represented the first encroachment upon and transformation of the old rural world – still without too dramatic effects – creating rather small urban settlements interspersed in the countryside. Starting before the war, but gaining momentum in the postwar years, a new organization of the industrial process took place. This change was signified by a growing benefit of scale and further processing, and, as shown by Ahnström (1982), by a growth in industrial services, thereby precipitating a more concentrated and centralized growth, particularly in the larger city regions and in some specialized, fairly central industrial areas.

However, by far the largest areas of the countries, containing a large share of the population, could be designated rural both morphologically and in functional linkages. There were, moreover, for some time at least, signs of counterurbanization in Denmark, Norway and Sweden, although restricted in scope to the areas near the cities (Illeris 1980, Ahnström 1986). This process was counterbalanced in the 1970s by persistent out-migration from the extreme periphery. This movement is spreading today. In Norway the majority of rural municipalities are experiencing net out-migration.

Based on economic structure and on later eligibility for special regional political measures, areas can be singled out which, although increasingly urban in settlement pattern, still exhibit a structure which justifies the term 'rural'. They generally coincide with the local communities towards which general and specific policies and plans have been directed to cope with growing economic and social disparities.

The maps in Figure 4.1 give only a rough approximation of this rural world. Some of the areas in central and southern Sweden so designated are undoubtedly rural, but are not eligible for regional assistance. Conversely, there are in the development areas of northern Norway some few medium-sized towns which are unmistakably part of the urban world.

Norway and Sweden – Scandinavia in its most restricted meaning – are the most interesting and meaningful countries to compare for structural and political reasons. In order to emphasize issues and types of policy, only occasional references will be made to the other countries.

Rural development problems: inherent and man-made

Regional disparities, both at the geographical and the household level, have existed far back in history in all Scandinavian countries. The problems relate both to differences in income and to the provision of private and public services. Some of these problems are geographically conditioned, inherent in, and shaped by, the physical environment. Climate, topography and distance are the main determinants in this context, leaving their impact both on industrial and employment potential and on the costs of providing basic infrastructure and services. To these must be added the problems of adjustment in primary industries: agriculture, forestry and fisheries, which sprang up particularly in the aftermath of World War II. These were later supplemented with the vulnerability of hosts of primary processing industrial communities, owing to an ever-increasing acceleration of industrial restructuring.

The latter process precipitated demographic problems, with loss of population and distortion of the age and sex structure.

The infrastructural and service problems have always been evident, partly due to the existence of the vast, sparsely populated areas of, particularly, Norway, Sweden and Finland. They loom large as far as communications and social services are concerned. In Norway as late as the

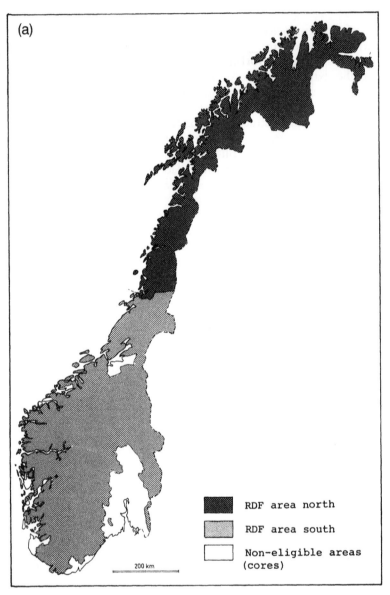

(a)

RDF area north

RDF area south

Non-eligible areas
(cores)

200 km

Figure 4.1 (a) Development areas (hatched) and urban regions (white) in Norway. (b) Rural and urban areas by municipalities in Sweden 1984. Extensive urban municipalities distort the real built-up area pattern, particularly in the northern part of the country.

Sources: (a) elaborated from maps by the Regional Development fund (RDF) (1980).
(b) elaborated from SOU (1984: 74).

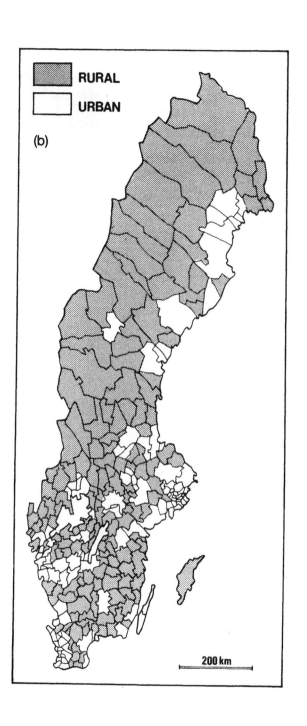

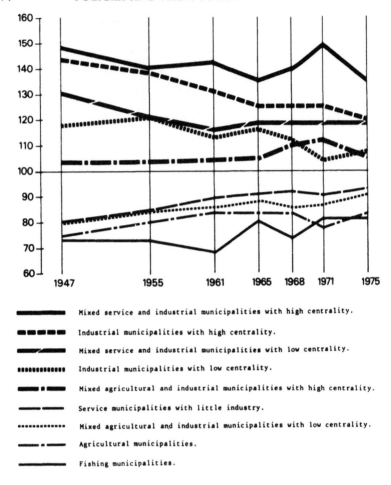

Figure 4.2 Relative development in growth of income per capita by municipal types. Country average = 100.

Source: Elaborated from Strand (1979).

1950s thousands of farmsteads and many fishing villages even on the mainland still lacked roads for motorized traffic. As late as 1979 there were, according to a survey (Strand 1979), 700 places without road or ferry links with the outside world. Even today there are, in places, important trunk roads, particularly in the western part of the country, without the

minimum standard of two single traffic lanes. As far as other communications are concerned, telephone automation was an exception in rural
areas through the 1950s, and was delayed for a considerable time. Only in
the mid-1980s is this minimum standard requirement shared throughout
the country by rural and urban areas alike. Time lags also abound in the
development of the radio and TV network in sparsely populated areas.

These problems have been common to most Norwegian rural regions,
regardless of location within the country, except for the eastern part,
because of relief. Similar problems in Sweden are confined to the vast
northern areas and are mainly due to distance and low population density.
These factors, combined with scarcity of personnel, also played an important rôle in creating disparities in social services, mainly education and
health services, particularly in North Norway. But in Sweden, too, the
supply of services was extremely scattered. Thus, in the regions north of
Sundsvall, covering about half the surface of the country, there were as late
as 1950 only 11 secondary schools (Ahlmann 1950).

There has always been a problem linked to income and other levels of
living conditions, much associated with the industrial structure. This is
clearly shown in Figure 4.2 for Norway and Figure 4.3 for Sweden, which
disclose in both countries in the late 1940s low income as a particular
problem for areas mainly dependent on primary industries. Only in the
central rural areas, or where natural resources for semi-processing industry
were close at hand, was there an income level approximating the average
for the country.

Rural problems due to loss of population and distortion of the population structure were originally the result of a rationalization process in the
primary industries, mainly in agriculture.

As shown in Table 4.1, there has been a continuous shedding of labour in
primary industries since the war. This development is common to all Scandinavian countries, though least pronounced in Finland during the first
decade, where resettlement of nearly half a million Karelians meant considerable colonization of new lands.

The considerable manpower losses in agriculture were not only a passive
adjustment to national economic development but were the result of a
vigorous specialization, sponsored by farmers and their organization alike
(Haga 1978). To this was added the problem of readjustment in primary
processing industry. This has been a special rural issue both in Sweden and
Norway, as many of the small industrial communities are close to raw
material and energy sources (minerals, metals and waterfalls), often in
remote places which lack potential as service centres for a wider
hinterland.

The problems of growth and repartition, which were not new to rural

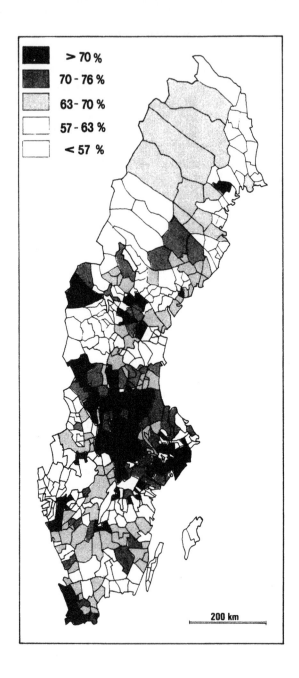

Figure 4.3 Per capita income by municipalities in rural areas in Sweden (1949) as percentage of average income in the whole country.

Source: Elaborated from Sverige Nu (1950).

Table 4.1 Economically active population in primary industries in Scandinavia, 1950–80, in absolute numbers (1000) and percentages.

	1950 Absolute numbers	%	1960 Absolute numbers	%	1970 Absolute numbers	%	1980 Absolute numbers	%
Denmark	519	25.2	366	17.8	244	10.6	201	7.7
Finland	912	45.9	721	35.5	429	20.2	279	12.6
Norway	360	26.4	274	19.5	170	11.6	105	7.6
Sweden	632	20.4	474	13.8	277	8.1	206	5.5

Source: RDF (Distrikteres Utbyggingsfond) 1980.

areas, thus became fairly general by the 1950s. Instruments were not developed, however, to cope with the problems, policies and planning creating new problems rather than redressing old ones.

According to prevailing views, the problems were abated through maximum adjustment to modern capitalistic development, which might as easily entail transfer of labour from the periphery as provision of industries there. In Norway, where the government was strongly committed to macroeconomic planning, whose efficacy in contributing to development has been questioned (Bergh 1978), projects in rural areas were mostly promoted as part of a growth strategy for the whole economy. The North-Norway plan (1952–60), the first regional plan in Scandinavia, was motivated in this way (Wood 1965).

In Sweden the strong central planning institution, the Labour Market Board (Arbetsmarknadsstyrelsen), and from the late 1940s the local physical planning authorities, were used systematically as instruments in facilitating labour mobility from rural problem areas. Simultaneously, the municipal planning institutions helped in easing this mobility, seeing that new infrastructure was provided to accommodate this labour force in the centres of growth.

The legendary Swedish welfare policy was already being formed; it legitimized the transfer of rural people to more concentrated settlements. It was believed that if industry were concentrated employees would, in the long run, cease to be affected by industrial transition problems. Complex industrial environments would be able to absorb shocks. Access to public and private services would, furthermore, be facilitated through this mobility and disparities would tend to disappear.

The outcome in Norway was a considerably lower concentration of activity and population than in Sweden: this was partly due to a different structural development, entailing a lower degree of industrial refinement and greater concentration on the processing of raw materials, which were more peripherally located. Partly, however, the difference was due to a less systematic policy of labour transfer. The only direct state intervention into mobility in Norway concerned transfer of people from remote islands. This policy was pursued vigorously, and, between 1946 and 1971, it helped to reduce the number of settled islands from 2100 to 1200.

Finally, a stronger resistance to centralization was built into the geographical structure of the country, coupled with some opposition to specialization in primary industries, and a persistence of the old integrated household economy, as pointed out by Brox (1984) for the northern periphery.

The new orientation: regional policies and plans as a development from above

THE CHANGE IN POLITICAL CLIMATE: POLITICAL ORGANIZATION AS A BASIS FOR RURAL POLICIES AND PLANS

By the early 1960s the redistribution of industries and population from peripheral areas began to cause concern in wide circles in all Scandinavian countries, including in Sweden. This concern was partly economically motivated. On the one hand, there was the growing demand for both physical and social infrastructure, coupled with increasing planning and implementation costs. On the other, existing infrastructure, some of which had recently been developed, was in danger of being not only suboptimally used but entirely lost in the out-migration areas. Strategic considerations also played some part in this new concern, particularly in the peripheral northern parts of the countries where maintenance of a certain basic population was deemed necessary for defence purposes.

Most important, however, was a reorientation of the sociopolitical climate towards a preoccupation with rural problems and rural values, although it was not until the 1970s that 'green' ideas became prominent in politics and began permeating most political parties. The Scandinavian countries, particularly Finland and Norway, were still first-generation urbanized societies. And, what is more, rural interests were well organized.

In Norway these interests wielded great influence not only in the liberal and centre parties but in the Social Democratic party as well. In Sweden the Centre party developed a more independent political platform than in the first postwar decade. Although it had its stronghold in rural areas, it began attracting more followers than before even in urban areas and started from the 1960s to demand a reorientation in economic policy towards a more balanced regional growth. In Denmark likewise the original liberal party (Venstre) had its stronghold in the countryside.

Simultaneously, the traditional political and institutional set up at lower administrative levels favoured a more decentralized, rural value orientation. This was particularly the case in Norway where the municipalities, through their councils, had been important political institutions since the 1830s, originally introduced by the farmer majority in Parliament as a defence against an arbitrary display of power by central authorities in the local communities. The right to raise money, even levying income tax, emphasized the independent position of the lowest political level, which, in reality, also organized the feeble local democracy at county level. The local government platform was of shorter duration in Sweden in independence and scope. In both countries, however, a reorganization of the political institutions of the rural areas took place in the postwar period, first in Sweden where the number of municipalities had been reduced from more than 2500 to about 1000 by the early 1950s. In Norway the reorganization through amalgamation and other changes in frontiers was implemented by parliamentary decision by the mid-1960s, reducing the number of units from about 750 to about 440. By that time provision had already been made in Sweden for an even more drastic cut in the number of units – down to about 270. The new and larger municipalities were believed to create more efficient service institutions for the population by strengthening the financial basis and planning capacity of the local political and administrative level. Although the reorganization was motivated by a better foundation for independent action, the new administration later turned out to become mainly an agent for implementing administrative routines and service provision determined at the state level. This was particularly the case with compulsory education and social welfare. Thus, both gain and loss have been attributed to the administrative reorganization, the loss being mainly in political participation by local people. This has prompted a discussion of the need for smaller planning and decision-making units within the municipalities. Also the introduction of the municipal and regional planning institution in Norway in 1965, the counterpart of which had been practised in Sweden from the early 1950s, had been a decree from above. By this measure, for which many rural municipal councils proved to be ill prepared, more orderly physical and economic organization and more adequate priorities for unrestricted use of resources were believed to be promoted. In reality it proved difficult to consider all the conflicting needs in an ever more heterogeneous countryside (Kyllingstad 1984). Moreover, great inequalities persisted and a need for co-ordinating the use of resources at an intermediate level also arose.

To solve this problem a new county organization was introduced in Norway in 1974, providing for directly elected county councils. Among other tasks, the new organization was to plan, in a more orderly fashion, the development and maintenance of higher education, hospitals and communications.

This was partly modelled on the Danish county organization, which was instituted at the same time. In Sweden the county board persisted as the long arm of the state, supplied with representatives from the municipalities (Länsstyrelsen) into the 1980s, when a functionally limited, directly elected body, Landstinget, was introduced.

GOALS AND PRINCIPLES OF REGIONAL POLICY

The systems outlined above, which due to the space available are sketched very broadly, are important instruments for channelling policies and plans, aiming at levelling rural disparities and redistributing growth – what in lack of a better term is labelled 'regional policy'.

The principles of these policies and plans, which in all Scandinavian countries have mainly come to be concerned not only with rural but even with peripheral rural areas, were not formulated once and for all. They are the results of a political process working over time. Goals were first explicitly formulated in Norway and were much linked to the objective of maintaining the settlement pattern in its main features. To this was added, explicitly in the early 1970s, the principle of promoting a uniform level of living, entailing a maximum distribution of public and private services. Later, particularly as a reflection of the 'green' wave mentioned earlier, a concern for resource utilization and environmental protection was added to the list.

The objectives of the Swedish regional policies, first formulated by Parliament (the Riksdag) in 1965, are not fundamentally different, but there are clear nuances. In accordance with the emphasis on social equity, the Swedish goals have been expressed more in welfare political terms, giving prominence to work, service and a good environment as essentials (SOU 1984: 74). Also, in the guidelines for Swedish regional policy there is concern about the settlement pattern (Wärneryd 1984). But whereas in Norway the preoccupation has been with the lowest level of settlement, explicitly formulated in a White Paper in 1973 (Authoritative Parliamentary Report No. 50 1972–73, St. Meld. 50), the Swedes have approached the problem more from above by stating that the main objective is to pro-mote a balanced development of size and structure of population between the main regions of the country. This means that in Sweden the develop-ment of strong regional centres has been overtly encouraged as a part of regional policy, designating them support points (*stödjepunkt*). A hierarchy of central places has even been drawn up, from big-city alternatives through regional growth centres down to a third level of service centres and other centres, to serve as active agents in rural socio-economic development.

As far as the *details* of the settlement pattern are concerned, in Norway there has been no consistent setting of goals. Attempts have even been made, at some instigation from above, to concentrate the population in the countryside. Following the idea of growth poles, nine centres were selec-ted in 1965 in different parts of the country, thought to serve as monitors of regional development in their respective regions. However, this move by a social democratic government was counterbalanced the following year by a centre-right government, through the creation of six so-called develop-ment regions. This was a move intended to experiment with the promotion of growth at the lowest possible level.

Later settlement goals have more or less come to converge in the two countries as the promotion of the rigid centre pattern has not been pushed any further in Sweden. In both countries during the last ten years it has been left to the new county institutions to formulate the goals and policies of the settlement system.

Regional policy, which by the late 1970s was downgraded in Sweden in relation to the national goal of full employment (Stenstadvold 1981), has been based on the principle of a 'sound market economy' in both countries. Thus the fundamental goal formulated in the preamble to the Norwegian Regional Development Fund is the creation of durable, profitable places of work. This means, however, that there are conflicting political goals and objectives. The profitability and efficiency of enterprises and industries have generally presupposed capital intensity and centralization. Regional stability needs a smaller-scale, decentralized industrial structure.

PLANNING TOOLS AND REGIONAL MEASURES

In order to realize the goals outlined above, a system of direct and indirect measures has been developed, generally *ad hoc* in a long development pro-cess. We may in this context distinguish between broad development plan-ning at different levels and specific and direct use of different support instruments, although the two may be closely interrelated. Furthermore, a distinction is relevant between policies and instruments for promoting single industries and institutions (sector policies) and sector-overriding measures directed towards broader development in a territorial context (regional policies). In this section we will first examine the measures and tools of the last-mentioned type.

Irrespective of how goals are formulated in settlement or employment terms, long-range development is dependent on industrial growth of some magnitude. There are many ways of channelling and promoting industrial development. Fundamental to the regional policy which was pursued up to at least the mid-1970s was the notion of redistribution of activity. This signified provision by economic incentives for movement of industrial enterprises from growth areas towards the periphery either by total relocation or by branch plants. This method of regional growth leaned in its promotion heavily towards manufacturing industries, but was, especially in Sweden, supplemented with the relocation of non-administrative public institutions from the capital to more peripheral regional centres.

Simultaneously, industrial efforts originating in the rural areas were sought, taken care of and incorporated into the support system being developed. This was particularly the case in Norway, where institutions for sponsoring local initiative and channelling activities from the outside were created quite early (county boards of employment and local initiative).

As pointed out above, when considering industrial promotion work it is fruitful to distinguish between general policies and plans and the more specific support measures which are directed towards enterprises. The former, mainly conditions of regional measures and support for infrastructure, have in both countries become incorporated into the political-institutional hierarchy. In Norway they have at the central level been taken care of by two ministries, initially by the Department of Labour and Local Government, from the 1970s also by the Department of Environment. These institutions have provided for the development of industrial estates and have given support to municipalities in building premises for lease and undertaking other basic investments. In Sweden it has been the task of the Department of Industry to initiate, co-ordinate and, to a certain extent, implement different types of support measures according to existing regulations.

Delegation and even devolution to lower political and administrative levels have since become fairly common. In Sweden a county planning scheme (*Länsplanering*) by the county boards was started in 1967 and was intended to become a regular procedure for deciding priorities on important regional issues. After completing two planning rounds, each covering four years on employment, industrial and settlement issues, this planning scheme, strongly influenced from above, was later modified to a less formal procedure.

A parallel county planning scheme was instituted in Norway in 1973. Originally very broad in scope, covering almost any regional development issue, it later became concerned mainly with policy areas specific to educational, health and communication issues.

This has undoubtedly increased competence in rural areas through the development of a more decentralized secondary education system. However, the county planning instrument has been less successful in both countries in co-ordinating important sectoral planning policies into more territorially based development instruments.

The basic instrument of industrial growth has always been the single firm or institution. Regional policy has, therefore, mainly been an industrial policy, with the objective of increasing the range of activities to compensate for lost employment, originally in agriculture but increasingly in semi-processing manufacture. The idea behind the support measures is then, to compensate for higher costs of development in what are considered unfavourable locations.

The Norwegian system grew out of the North Norway Development fund which was mentioned above. It was reorganized as a Regional Development Fund (RDF) for the whole country in 1961. Originally geographical eligibility was unsystematic, but in 1971 development areas were designated, for which tax-exempt funds might be set aside by industries (*Utbyggingsområder etter Distriktsskatteloven*). These areas, the delimitation of which was based on development problem indicators – unemployment migration, depopulation and cost disadvantages for location – were ultimately determined by political bargaining and have later been subject to minor reorganization. The area in 1980 is shown in Figure 4.1a. Reflecting the prevailing ideas of strategic growth components of the 1960s, only manufacturing industry was originally eligible for support. The range of eligible activities has since become extended, but eligibility is still mainly restricted to industries characterized by an export base, such as wholesaling, tourism and business services.

The Swedish system, which was organized later than its Norwegian counterpart, does not differ from it substantially, but there are subtle distinctions. Funds for assistance are mostly administered centrally by the Department of Industry. The various decentralized regional development funds in the counties are also highly dependent on allowances from the state, and are much smaller than the central funds. Also, in Sweden geographical eligibility has changed over time. It was originally more restricted to the northern peripheral regions than is the case in Norway (Fig. 4.4a), but substantial areas are covered. Later it was differentiated into more graded areas of support, A, B and C areas (Fig. 4.4b), a counterpart of which has also evolved in Norway, based on intensity of investment grants (Fig. 4.5).

As far as types of eligible industries are concerned, only small differences exist between the countries.

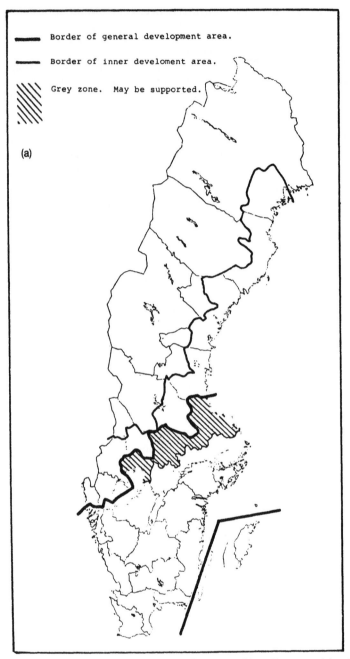

Figure 4.4 (a) Development areas in Sweden, 1975. (b) Differentiated development areas in Sweden by municipalities 1984.

Source: SOU (1984: 74).

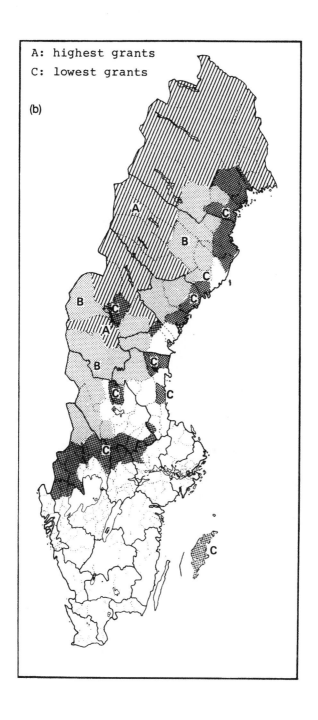

A: highest grants
C: lowest grants

(b)

Investment grants

☐ NO SUPPORT
▨ 15% SUPPORT
▩ 25% SUPPORT
■ 35% SUPPORT

Source: Regional Deve-
lopment Fund 1982.

Figure 4.5 Differentiated development areas and non-assisted areas in Norway.

Source: Regional Development Fund (1982).

Grants are selective in all countries, based on application from the firms themselves. The Norwegian system has been the most decentralized from the outset, as the County Board of Employment and Local Initiative has acted as an advisory body for the RDF. Thus, some local priority decisions may be provided for. From 1982, grants up to a certain amount can even be given final approval in the new County Board for Industry. In Sweden, only grants from the county funds can be approved at that level. Some decentralization has also been instituted in central funding, however, by vesting in the counties the right of allocating priorities for spending of allowances.

Table 4.2 shows the main types of support measures available in the two countries by the 1980s for firms in eligible industries, extended for the sake of comparison to Denmark and Finland. From the figure it can be seen that three main types of assistance have been practised, though somewhat differently in the different countries. These are (1) financial instruments, which are also the initial ones, covering a wide range of measures, (2) employment subsidies, and (3) locational control. Norway has developed the broadest range of instruments, but also the most one-sided, relying mainly on capital subsidies, of which cheap loans for top financing were the original measures. Later, loans were replaced by investment grants covering nearly all types of productive capital input but restricted to 35% of the total investment value.

From the 1960s a system of transport subsidies was introduced: they most benefited firms in the peripheral northern and western parts of the country. As far as subsidies to labour are concerned, Norway has no direct grant for job creation, but development areas have, since the late 1970s, enjoyed some tax exemptions on labour. As the only Scandinavian country, Norway in 1977 introduced a system of establishment control, built into

Table 4.2 Some regional political measures in the Nordic countries.

Countries

Measures	Denmark	Finland	Norway	Sweden
investment grants	*	*	*	*
subsidized loans	*	*	*	*
tax exemptions		*	*	*
job-creation subsidies		*		*
differentiated labour taxation			*	
transport subsidies	*	*	*	*
establishment control			*	

Source: The Nordic Official Reports Series (NU) 1978.

the RDF system and intended as a powerful instrument to reinforce redistribution of investments from central to peripheral areas. However, the instrument has fallen short of expectations, as it was introduced at a time when investment activities were generally low.

Unlike Norway, Sweden has introduced a system of premiums for job creation which is effective over several years, differentiated according to the graded areas mentioned above. As in Norway, financial measures are the most widespread and substantial, as can be gathered from Table 4.3

According to the figures, investment subsidies and different types of cheap loans constitute the main part of regional support in both countries, the latter being more predominant in Norway. Sweden leads in absolute size of support given, and Norway in per capita spending in the development areas, even if we allow for some higher real exchange value of the Swedish krone than is shown in the table.

SECTOR MEASURES AND INSTRUMENTS

The picture of industrial support for regional development is incomplete if the transfer of resources channelled through the strong sectorial bodies at central and intermediate levels is not taken into consideration. Primarily this is a support for agriculture, fisheries, manufacturing industry and a few services. Its main objective is to keep enterprises viable and retain employment, but it also acts as an instrument for income redistribution.

There are no comparative statistics for direct regional political versus sectoral spending in development areas in the different countries. An estimate for Norway for 1980 and 1982 emphasizes the growing predominance of sectoral measures (Table 4.4). Support to primary industries

Table 4.3   Direct regional political support to industry in Norway and Sweden (Mio.kr.).

	1982 Norway (1 N.kr., approx. £0.105)	1982–3 Sweden (1 Sw.kr., approx. £0.100)
regional transport subsidies	121	181
investment and other financial grants	365	554
loans and guarantees (costs)	275	158
other (including employment subsidies)	54	119
Total	815	1012

Sources: NOU (1984) 21A; SOU (1984:74).

Table 4.4 Public spending on industries in development areas in Norway, 1980 and 1982 (Mio. N.kr.).

	1980			1982		
	Sectorial transfer	Regional political transfer by the RDF	Total	Sectorial transfer	Regional political transfer by the RDF	Total
agriculture	4174		4174	6184		6184
thereof regionally differentiated	1593			2447		
forestry	120		120	124		124
fisheries	894		894	956		956
manufacturing	838	524	1362	2480	695	3117
private services	185	81	266	258	120	378
total	6211	605	6816	10002	815	10817

Source: NOU (1984) 21A.

looms particularly large, but manufacturing industry receives a growing share. The tendency is probably the same in Sweden (SOU 1984: 74) but not as pronouncedly in favour of sectoral measures.

Given partly on non-territorial criteria, this support may distort the broad objectives of regional development policies. Agricultural subsidies have thus come under attack in the regional debate in Norway, an issue to which we shall return below. However, there is disagreement as to the effects. Brox (1980), in his general critique of Norwegian regional policy, has nevertheless credited the uncoordinated sector policies with some positive regional effects.

In both countries the social security and welfare systems are the main agents of transfer of resources. This redistribution is client oriented and general in nature, based on rules and regulations. It not only contributes to pumping income and purchasing power into local communities but also creates considerable employment. In its impact it is regionally biased, as peripheral rural regions with their lower income generation may profit relatively more from redistribution of social and welfare benefits than more central areas. Authorities may, furthermore, be tempted to use social instruments of this kind instead of more active measures – for example, in order to fight unemployment. There are no statistics throwing light on the importance of transfer of welfare, viewed as a regional measure in the Nordic countries. Per capita public consumption in health care and welfare is particularly high in Sweden, compared with Finland and Norway. Redistribution has the highest social profile in Sweden, and this might indicate

that the most substantial regional effects are to be found there. It is probably one of the main reasons for employment and population stabilization
in the Swedish peripheral regions during the 1970s. Identical results may be
found in Norway, but they are not so pronounced (Sjøholt 1984).

Rural stabilization through these measures is a typical example of planning from above, but in a rather haphazard way. The further potential of
the instrument in rural and regional development policies has been
strongly questioned in recent years, both for ideological and economic
reasons.

Results of regional policies and regional planning

The task of analysing and evaluating the effects of rural policies and plans
may be approached from different angles, according to the established
goals. Direct employment effect is one factor, improvement in the robustness of industrial structure another (particularly important for future performance) and settlement, demographic and welfare impact a third.
Assessment is, however, highly problematic. Although performance,
according to these indicators, is possible to establish over time, it is more
than difficult to isolate the effects due to particular regional policies from
those which have accrued from sectoral policies or are the result of
spontaneous development. Sectoral policies have undoubtedly had far-
reaching impacts. In Norway, Parliament decided in 1975 to bring average
incomes in agriculture over a short span of years on par with average wages
in manufacturing industry, the only recorded example of this policy in any
OECD country. This move produced multiplier effects which had an
impact of consolidation on some rural areas, although they were negligible
in the northern periphery. But a number of unintended consequences
followed in its wake, such as biased repartition and, in conjunction with
that, too heavy capital investments which in turn tended to offset some of
the consolidation effects. Moreover, it produced a surplus in milk and
mutton. In other industries there are, from Swedish research (SOU 1984:
74), indications of improved robustness in rural areas during the last 15
years by better adjustment to technology, more product innovations and
higher productivity. Simultaneously, the structure of manufacturing
changed to a higher share of typical growth industries in the development
areas. However, this must not be taken as a conclusive argument for the
positive contribution of regional support measures. It may well be that
investment would have followed in any case. In a micro approach to the
problem, Fredriksson & Lindmark (1978) claim that there are clear limits to
the effect of financial regional measures. Capital subventions generally
give cost benefits to capital intensive enterprises only. The support for job

creation, however, has contributed to better performance in a wide spectre of enterprises.

In Norway it has been borne out from a recent investigation (Strøm 1984) that leaders of new enterprises attach great significance to the contributions by the RDF in overcoming establishment thresholds.

In Swedish manufacturing industry as a whole the peripheral areas witnessed relative growth up to 1980, whereas most other parts of the country suffered a decline. Simultaneously, there was manifest growth in private services in earlier problem counties. The same trend is evident in Norway. Shift-share analysis showed a clear redistribution of manufacturing employment to the development areas, and within these to the rural areas (Isaksen 1984, Bivand 1984). Distinguishing between trends in periods with and without regional political instruments in operation, Bivand (1985) was led to conclude that, during the 1970s, about 1000 new jobs per year were created as a result of regional measures.

This growth only partially met the need in the countryside. After the old household economy with its traditional sex rôles was torn apart, jobs for women became increasingly scarce in rural areas throughout Scandinavia. With growing aspirations and competence among women, this became a source of conflict. The imbalance was but little offset by the new industrial structure. Some improvement took place during the 1970s, as service, mainly women's jobs, gave positive employment transfers to rural areas –in Norway at any rate up to 1981 (Sjøholt 1984). This mainly benefited people in the new rural centres, where most of the jobs were located. The growth of these nodes was largely dependent on public service employment, making services the new rural industry. People in contracting, sparsely populated areas were made increasingly dependent on commuting, which placed them at a disadvantage, especially the women. The same tendencies were found in northern Sweden (Björnberg & Dahlgren 1984).

The effect on the main settlement pattern, the central Norwegian regional political goal, is difficult to measure and is highly controversial. Brox (1980) claims consolidation in rural settlement, based on data from North Norway, attributing it to a low employment participation rate, excessive commuting and transfer of income from the state. Illeris (1980) and Ahnström (1986) infer a general counterurbanization trend, viewed in a Scandinavian context, the former giving regional policy most of the credit for the development. Although there has been evidence of some population stabilization in rural, non-central regions, this is mostly the case in the strongholds of the areas. That was the conclusion of a Swedish analysis (SOU: 74). That consolidation is far from being a universal trend was shown by Hansen (1982) with empirical data from coastal districts in North Norway. The northernmost Norwegian county, Finnmark, has lost population over the last ten years. From 1981 there are clear signs of growing concentration and centralization of population in most Nordic countries, a return to the capital city regions.

Carefully interpreted, there is in the late 1980s a tendency towards strengthening of rural communities in central areas and continued contraction of population and settlement in the extreme periphery throughout Scandinavia. In Norway some growth in manufacturing in, and redistribution of, service jobs to the latter areas has barely compensated for the loss in primary occupations. Some of the new industry is also fragile in the sense that its resource base is exogenous rather than endogenous. Some districts have profited from the petroleum activity, but mostly in semi-skilled jobs on the installations in the North Sea and only occasionally in on-site industrial activity. Moreover, the increasing flagging out of the merchant marine has deprived rural Norway of many former good and safe jobs. In cases of net gain of jobs, as reported from the Swedish periphery, this has not led to population growth. Most of the new jobs are for married women, often in part-time work. The demographic structure of these areas is vulnerable, deaths exceeding births in an increasing number of municipalities, casting doubt on potential future consolidation (Hansen 1986).

As to redistribution of welfare, to a large extent the result of political action, the range in income has decreased in all Nordic countries during the postwar period (clearly demonstrated for Norway in Fig. 4.2). According to Oscarsson & Öberg (1987), no county average in income deviated more from the national mean than 20% in Finland and 15% in Norway and Sweden in 1980. Analysis of a broad spectre of life-quality components indicate, though, a persistent imbalance in the extreme periphery, as testified by studies in Finnmark, Norway (Aase 1986, Lindkvist 1986).

Rural policies and planning on the threshold of the 1990s: some new tendencies

Organization and instruments for regional policy were developed in a period of general industrial growth and were in their means adapted to redistribution from above of surplus generated in this process. In the sluggish growth of the economy from the mid-1970s, now also manifested by increasing turbulence in highly specialized secondary industry, this policy has increasingly come into conflict with legitimate demands raised also in more central areas. And, what is more, the regional policy itself, not only its instruments and organization but even its goals, has become an object of criticism and attack (Brox 1984).

New growth industries have other needs than the traditional ones and are characterized by small-scale innovations, more flexibility in production systems and the ability to readjust. In their development and consolidation they are increasingly dependent on human capital and competence, whereas heavy capital equipment used to be the main factor in realizing mass production through benefits of scale. There is also a need for new types of sophisticated services based on information technology. This development calls for new kinds of entrepreneurs and represents a challenge for rural planning and policies. There are in rural areas no ready comparative advantages for this reorientation. A laborious process of competence development is necessary for their mobilization. The traditional institutions are inadequate to this end.

From the mid-1970s nearly all development aid taken up in Norway has been by local firms (Isaksen 1984). From the 1980s the country has witnessed many attempts at initiating and co-ordinating entrepreneurship at the local level by more unconventional means. This is, above all, characterized by more self-reliance than before, although promoters still have the function of channelling the instruments and measures available in the existing system to the benefit of the new entrepreneurs. In this process it is, as in other Western countries, the small enterprises which generally represent the innovation. They thrive best in active innovative environments which function as seedbeds for the development and spread of competence, rather than in milieux primarily investing in infrastructure for mobile branch plants. Although there are thresholds for this development to take off, such interactive environments may also come into being in rural areas. Numerous examples have been cited, among them Gnossjö in Southern Sweden, from which the concept 'Gnossjö spirit' is derived (Thörnquist 1986), and Osterøy in West Norway (Mjøs & Jevnaker 1979).

A systematic promotion along the new lines has mainly been adopted in Norway through the search for, and guidance of, entrepreneurs (bootstrap-pulling strategies), through broad local mobilizing of ideas and projects (idea banks) and, to a lesser extent, through local development corporations and direct municipal initiative. Originally this mobilization was initiated by non-political organizations or local politicians, search for entrepreneurs by a parastatal development board in West Norway, the idea bank by a municipality in the same area and the development corporation by local politicians in North Norway. These initiatives from below have been followed up by central development institutions. Thus a search and guidance programme for entrepreneurs implemented by the RDF in North Norway in the early 1980s claims the founding of about a hundred small enterprises. In its wake has followed the introduction of new types of support measures by the RDF, geared less towards financial aid than towards product development, management, marketing consultancy and general readjustment. For the financial year 1984 this new expenditure, approximately 180 mio. NOK (17 mio. £), made up nearly a fifth of the total engagements of the fund. Less weight has been given to automatic ear-marked transfers to municipalities from the state. From 1985 a system of block grants was introduced; they were intended to foster local priorities and initiative.

Similar reorientation is taking place in other Scandinavian countries, but it is less systematically applied. In Sweden, readjustment experiments have been promoted, jointly financed by state and counties as a response to crises in local industry. During the 1980s non-financial regional measures have been increasingly adopted. There are even signs of mobilization for the creation of new employment, linked to the new municipal organization at local community level, which was introduced in the early 1980s, as reported from the Norrland inland by Ström (1985).

Conclusion

When trying to sum up manifestations in rural policies and planning in Norway and Sweden on the threshold of the 1990s we meet with a complicated picture. The prospect of non-growth in industrial and social redistribution has enforced a systematic rethinking of regional policies and instruments. Special commissions have, in the 1980s, dealt with the issues in both countries. There is a lot of common ground in their recommendations, but also country-specific ideosyncracies.

Both commissions come forward with propositions for more geographically restricted eligibility for support but simultaneously for a broadened range of support measures, complementing the classical financial instruments with more aid for readjustment of the industrial structure. A proposal for a fundamental reorientation is put forward by the Norwegian commission (*Bygdeutvalget*) (NOU 1984). Geographical differentiation of support should, according to the majority of the commission, be based on alternative cost of labour in different regions. The consequences of this proposition would be to give priority to regions whose alternative costs are low and reduce support where an alternative use of the labour force would give a higher return. The impact would be particularly great on repartition of support to agriculture, which would become linked less to production and more to the availability of alternative industrial opportunities.

In the Swedish recommendations, relative reductions of financial support loom large. Resources for enterprise development are to be concentrated even more on the real problem areas. There is, however, great concern about research and development functions and producer services and their impact on future regional development. Their promotion implies a strengthening of regional strongholds and may entail a conflict with the policy of decentralization and consolidation of peripheral local communities.

In a Norwegian Parliamentary Report (Authoritative Parliamentary Report No. 67, 1984–85, St. Meld. 67) the long-term need for developing more knowledge-based functions for the rural areas is emphasized. This might be accomplished by developing infrastructure for information networks as decentralized as possible in order to help peripheral areas to keep abreast in growth of information and knowledge-based industries. But this reorientation would be no panacea for a rural consolidation. A more sophisticated approach in fields where rural areas still have a comparative advantage, for instance fish farming in Norway and bio-energy in both countries, may perhaps give more realistic prospects for industrial development, at any rate for the rest of this century.

Viewed in a long-term perspective, planning and policies will probably have to be linked to demographic factors. In keeping with the conclusion in the last chapter, it is doubtful that rural Scandinavia will be able to maintain its population without considerable in-migration. In a situation with keen competition for human capital, such in-migration may be highly problematic to realize. This fact, therefore, will again require a reorientation: learning to plan for an ever sparser rural human environment.

References

Aase, A. 1986. Ten years of welfare geography. In *Welfare and Environment*, M. Jones (ed.), Trondheim: Tapir, 33–62.

Ahlmann, H.W. 1950. *Sverige nu Atlas över folk, land och näringar*. A.V. Carlsons Bokförlags AB, Stockholm.

Anhström, L. 1982. The concentration of a compound. The deconcentration of its parts. The economically active population of the Stockholm region 1950–1975. *Geografiska Annaler* **64** (2), 69–75.

Ahnström, L. 1986. The turnaround trend and the economically active population of seven capital regions in western Europe. *Norwegian Journal of Geography* **40** (2), 55–64.

Authoritative Parliamentary Report 1972–3. No. 50 (St. Meld. 50). *Om regionalpolitikken og lands- og landsdelsplanleggingen*. Oslo: Ministry of the Environment.

Authoritative Parliamentary Report 1984–5. No. 67 (St. Meld. 67). *Regional planlegging og distriktspolitikk*. Oslo: Ministry of Labour and Local Government.

Bergh, T. 1978. Ideals and realities in Norwegian macro-economic planning 1940–1965. *Scandinavian Journal of History*, 75–104.

Bivand, R. 1984. Endringer i industriens regionale sysselsettingsstruktur i Norge 1951–1981 og effekter av distriktspolitiske virkemidler. In *NOU 1984 , 21 B*, 226–37. Oslo: Ministry of Finance.

Bivand, R. 1985. The evaluation of Norwegian regional policy: parameter variation in regional shift models. *Environment and Planning C: Government and Policy* **4**, 71–90.

Björnberg, U. & L. Dahlgren 1984. Balanser och obalanser i Sveriges regionala utveckling. *NORDREFO* **15**(3), 55–86.

Brox, O. 1980. Mot et konsolidert bosettingsmønster? *Tidsskrift for samfunnsforskning* 3–4, 227–244.

Brox, O. 1982. Fem forsøk på å planlegge Nord-Norge. In *Planleggingens muligheter. (2) Forvaltning av regionene*, N. Veggeland (ed.), Universitetsforlaget Oslo. 13–44.

Brox, O. 1984. Økonomisk og velferdsmessig forankring av regionalpolitikken. *NORDREFO* **15**(3), 13–33.

Fredriksson, C. & L. Lindmark 1978. Företagen och regionalpolitiken. *NORDREFO*. Stockholm: Ministry of Industry.

Haga, G. 1978. *Avviklingsbonden og hans representantar*. Oslo: Cultura.

Hansen, J.C. 1982. Bosettingsmønster i utkant: Statisk eller dynamisk. In *Folkemakt og regional utvikling*, R. Nilsen, J.E. Reiersen & N. Aarsaether (eds), Oslo: Pax, 117–129.

Hansen, J.C. 1986. The urban turnaround – the beginning of a U-turn? In *Demographic Issues: Migration and Ethnic Minorities*, M. Cawley (ed), 8th International Seminar on Marginal Regions in Association with University College Galway. 5–35.

Illeris, S. 1980. Regional koncentration ophörer i Norden. *NORDREFO* **11**(3–4), 23–41.

Isaksen, A. 1984. Industriarbeidsplasser i Bygde-Norge. Hva skjedde i 1970-åra. In *NOU* 1984:21B, 207–25. Oslo: Department of Finance.

Kyllingstad, R. 1984. Befolkningsnedgang som utfordring i samfunnsplanlegging. *Kart og plan* **45**(4), 491–2.

Lindkvist, K.B. 1986. *Levekår og naeringsutvikling i Finnmark*. Graduate thesis, Department of Geography, University of Bergen.

Mjøs, L. & B. Jevnaker 1979. *Samarbeid i Osterøy-industrien*. Arbeidsrapport nr. 22. Bergen: Industriøkonomisk Institutt.

Nordic Official Reports Series (NU) 1978. *Nordisk forskning om regionalpolitik i omvardling* A 1978:12.

Norwegian Official Reports (NOU) 1984:21 A *Statlig naeringsstøtte i distriktene. Bygdeutvalget*. Oslo: Ministry of Finance.

Oscarsson, G. & S. Öberg 1986. Northern Europe. In *Regional development in Western Europe*, H.D. Clout (ed.), Chichester: Wiley.

RDF (Distriktenes Utbyggingsfond) 1980. *Faglig og økonomisk bistand ved utbygging av naeringsvirksomhet*.

RDF (Distriktenes Utbyggingsfond) 1982. *Faglig og økonomisk bistand ved utbygging av naeringsvirksomhet*.

Sjøholt, P. 1984. The service system in Norway, viewed in the light of recent theory and socio-political development trends. *Norwegian Journal of Geography* **38**, 1–8.

Stenstadvold, K. 1981. Northern Europe. In *Regional development in Western Europe*, 2nd edn, H.D. Clout (ed.), Chichester: Wiley, 299–333.

Strand, S. 1979. *Ringvirkninger av transporttiltak. Norges vegløse steder*. Oslo: TØI.

Strand, T. 1979. *Økonomi og sysselsetjing i kommunane. Strukturtal og fordelingar*. Notat. Bergen: Maktutredningen.

Ström, Ø. 1984. DU-støtte og sysselsetting. In NOU 1984, 21B. 200–6.

Ström, L. 1985. *Local mobilization in Sweden*. Paper presented at 8th International Séminar on Marginal Regions, Galway, Ireland.

Swedish Official Reports (SOU) 1984:74 *Regional utveckling och mellanregional utjämning*. Stockholm: Ministry of Industry.

Thörnquist, G. 1986. Miljöer för förnyelse. *NORDREFO* **16** (1–2), 67–90.

Wood, J.S. 1965. *The North Norway plan: a study in regional economic development*. Bergen: Christian Michelsens Institutt.

Wärneryd, O. 1984. The Swedish national settlement system. In *Urbanization and Settlement Systems*, L.S. Bourne, R. Sinclair & K. Dziewonski (eds), Oxford: Oxford University Press, 92–112.

5   *France*

HUGH CLOUT

Rural policies and plans in context

France has an important legacy of historic towns and cities but mass
urbanization occurred rather more recently than in neighbouring countries
of North-West Europe (Pitte 1983). In 1950, 46% of the French population lived in the countryside and 28% of the labour force worked the land.
During the next two decades strong emphasis was placed on essentially
urban matters, such as establishing provincial growth poles and construct-
ing suburban housing to cope with the flood of urbanization that was under
way. But rural matters were not forgotten, with successive national plans
encouraging agricultural modernization and favouring the creation of
special rural development corporations. Very significant legislation was
passed in 1960 (the *Loi d'Orientation Agricole*) and 1962 (the *Loi Complémen-
taire*) to stimulate structural reform in agriculture and to enhance the living
conditions of country dwellers. As in earlier decades, cityward migration
and rural depopulation remained powerfully at work, raising profound
questions about the desirable distribution of population throughout the
nation (Berger 1975).

By the late 1960s other social trends were coming to the fore, with dispersed suburbia, retired migrants, rural tourism and decentralized
manufacturing giving rise to important new pressures in many but by no
means all country areas (Chavanes 1975, Hervieu 1978, Chevalier 1981).
Counterurbanization was to become more significant during the 1970s and
1980s, as was concern for environmental protection (Ogden 1985). Official
responses involved a spate of legislation which sought to conserve
cherished rural environments and to establish development plans for
'pressured' countrysides as well as for urban areas. The social and economic
problems of relatively remote rural districts were not forgotten and special
subsidies were made available in mountainous zones and other specified
areas. In the 1980s rural policy is no longer farm policy writ large, although
the fact that agricultural agencies continue to play a major rôle should not
be overlooked. Nor is rural management entirely a matter of 'top-down'
schemes transmitted from Paris to the far corners of the countryside.
Thinly populated *communes* have been encouraged to co-operate and use
their own initiative to generate development projects that would be eli-

gible for technical and financial support from a higher level (Biancarelli 1978).

The needs of rural dwellers – not just farmers – and the challenge of countryside management have attracted attention from all parts of the political spectrum. In addition to using the logic which attempts to present an objective case for rational management of resources throughout France (*l'aménagement du territoire*), it is clear that centre and right-of-centre political groups have supported rural issues in reaction to the left-wing electoral persuasion of many urban constituencies (House 1978). However, since 1981 members of the Socialist administration have also favoured schemes to promote 'national solidarity' and to enhance living standards throughout France. In so doing they have tried to break away from centuries of administrative centralization, thereby giving new powers to the 22 *régions* and accentuating the responsibilities of the 36 512 *communes* for managing their own resources. With nine-tenths of France still occupied by farmland and forests and 93% of its *communes* classified as 'rural', this trend has directed attention once again towards the countryside. Of course, some rural areas have had their social composition transformed by the presence of long-distance commuters, retired city folk and weekenders (Fig. 5.1a), but depopulation, and the associated loss of basic services, is still encountered in many parts of the nation (Duboscq & Mathieu 1985, Fruit 1985). Such problems remain particularly acute in *la France fragile*, which not only includes mountainous districts (Alps, Pyrenees, Massif Central) but also the eastern lowlands of the Paris Basin where the resident population falls below 20 persons/km^2 (Schéma Général d'Aménagement de la France 1981) (Fig. 5.2).

Finding one's way through the labyrinth now labelled '*l'aménagement rural*' is far from easy. Top-down, nationwide agricultural policies have spawned holistic but area-specific rural projects. Environmentally conscious schemes have attempted to reconcile conservation and development, and the central administration has given its blessing to community-based local initiatives. Scores of new agencies have been created, but old and powerful ministries (for example, Equipment and especially Agriculture, renamed Agriculture et Développement Rural in 1972) have been unwilling to relinquish responsibilities (De Farcy 1975). These overlapping themes can be combined in a multitude of ways, but an exploration of basic agricultural reforms, evolving area-based plans, and recent schemes for rural revival offers a manageable route.

Agricultural reforms

Two distinct types of agricultural reform have modifed the landscape, ecology and socio-economic composition of rural France in recent

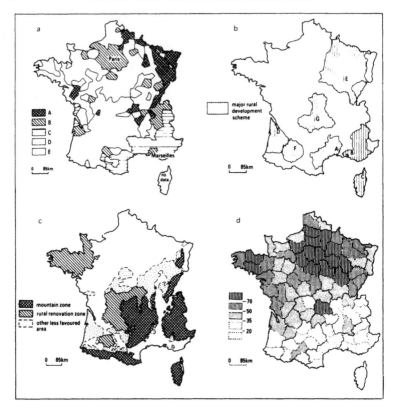

Figure 5.1 (a) Social composition of rural areas: A, industrial and/or working class; B, greater tertiary emphasis; C, tourism and/or retirement; D, strongly agricultural; E, close to average. (b) Mixed economy corporations: A, Bas-Rhône Languedoc; B, Canal de Provence; C, Corse; D, Landes de Gascogne; E, Friches et Taillis de l'Est; F, Coteaux de Gascogne; G, Auvergne-Limousin. (c) Less favoured rural areas. (d) Land consolidation (*remembrement*) completed 1982.

Sources: (a) after Fruit (1985). (b) after Brunet (1984).

decades. The first promoted consolidation (*remembrement*) of fragmented landownership units, and the second involved a package of measures to amalgamate and modernize functioning farms. In 1950 France had contained no fewer than 2 130 000 farm units (of more than 1 ha apiece), and many stretches of countryside were fragmented into tiny strips or blocks of land, with one man's holding intermixed with the property of his neighbours. The legacy of the past weighed heavily on rural France and

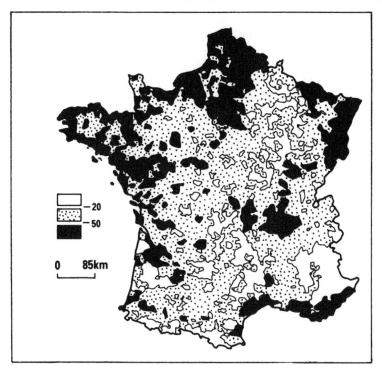

Figure 5.2   Population density 1982 (persons/km²).

conditions that may have been appropriate in the past posed massive draw-backs in the drive for postwar modernization.

Policies for refashioning the pattern of landholding (but *not* the size, number or layout of farms) had been introduced in 1918 to enable associ-ations of landowners in the war-torn northern *départements* to consolidate their holdings (Roudié 1983). Such opportunities were soon extended to the whole of France, but progress was slow and by 1939 only 385 000 ha of the 10 million ha deemed to be in need of *remembrement* had been treated. New legislation was passed in 1941 which widened the scope for action. Landowners, tenants, agricultural advisors or *préfets* could now initiate pro-posals which were scrutinized by special committees of owners and experts in each *commune*. If general agreement were reached, all plots were sur-veyed for size and quality and the land reallocated so that each owner received his fair share. Before 1963 one-fifth of the survey and reallocation costs had to be met by landowners, but now the state bears all direct costs.

With the help of generous loans, landowners still have to cover 30–80% of the cost of installing new farm roads, surfacing existing tracks, improving drainage, removing hedges and banks, and reclaiming abandoned plots. Many traditional farmers perceived *remembrement* as a 'revolutionary' change, and in terms of geometry and ecology it most certainly was, but although improvements in farm technique often followed, the process itself did nothing to increase the size of farms or reduce the number of farmers (Vergneau 1979). Nor did it involve any attempt to improve the quality or layout of rural housing.

All parts of France are eligible for *remembrement*, but landowners, agricultural advisers and local politicians have been particularly enthusiastic where the prospects for long-term improvement are great. At present, 12 million ha (40% of the farmed surface) have been consolidated, with three-quarters of all land being reparcelled in the rich and easily mechanized open fields of the Paris Basin, where there are many large, innovative holdings and where the costs of the operation are quite low (Fig. 5.1d). At the other extreme, relatively little has been achieved in upland France where environmental conditions are harsher, walls and hedges more abundant (thereby raising the cost of consolidation) and much terrain is not well suited to mechanized farming. Even so, some important schemes have been completed in the uplands as, for example, in the Puy-de-Dôme, where the political stimulus of Giscard d'Estaing should not be discounted. Throughout the country a further 10 million ha are judged to be in need of *remembrement*, sometimes for a second time.

The ecological implications of consolidation proved relatively modest across stretches of open field, but very different results occurred in areas with old, enclosed landscapes or with sloping terrain. For example, during the 1960s widespread consolidation was undertaken in Brittany where the hedges, banks, trackways and ditches of ancient *bocage* were removed on a vast scale and replaced by 'neo-openfields' (Flatrès 1979). As some agricultural experts feared, the ecological implications were disastrous, with soil being eroded by wind and water and accelerated runoff producing localized flooding (Deniel 1965, Rhun 1977). Removal of hedgerow habitats had serious effects on flora and fauna, with conditions being worsened by the use of large quantities of agro-chemicals. Vociferous protests followed and the activities of the newly formed Ministry for the Environment plus the passage of wide-reaching legislation on nature conservation (*Loi sur la Protection de la Nature*, 1976) led to an official reappraisal of *remembrement*. Environmental impact studies are now required and agricultural planners have to incorporate conservation features in their schemes. Individual consolidated zones are now far less extensive than they were 10–20 years ago, with a proportion of old hedgerows, banks and trees being retained. Special efforts are also made to plant new hedgerows and shelterbelts in vulnerable locations. Operation of the broader concept of *remembrement-aménagement* now means that zoning of land for future residential use, tourism and afforestation may be undertaken at the same time as plot consolidation (Anon. 1983). Construction of motorways and other major projects provide exceptional stimuli for stretches of *remembrement* to take place (Barre & Vaudois 1980) (Fig. 5.3).

The visual impact of consolidation is undeniable, but the second set of nationwide reforms have been more truly revolutionary. They originate in the convergence of official and professional viewpoints which led to innovative agricultural legislation in 1960 and 1962. The government insisted that farming needed to modernize if it were to fulfil its rôle in a rapidly industrializing nation and reap the benefits of membership of the EEC, and farmers' organizations, spanning a wide range of political tendencies, emphasized the need to improve the living standards of rural families. *Remembrement* was quite inadequate to promote the kinds of change that many young farmers had in mind (Houssel 1978). Old farmers needed encouragement to retire and sell their land so that farm enlargement could take place; and surplus agricultural labour needed to be directed into off-farm work. A battery of official measures was introduced to achieve these ends, with freshly released land being used to enlarge adjacent farms (but only up to carefully specified limits) or to install new farmers. (In fact, a scheme had operated since 1947 to facilitate young farmers to move from overcrowded agricultural regions such as Brittany in

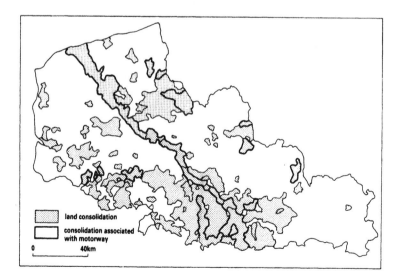

Figure 5.3 Land consolidation in the Nord.
Source: after Barre & Vaudois 1980.

order to settle in the depopulated Massif Central and South-West where farms were more readily available.) Agricultural training was improved as a result of the 1960–2 legislation and schemes were introduced to encourage farmers to share expensive equipment and to undertake other activities in common. As the prime allocator of financial support, the Fonds d'Action Sociale pour l'Aménagement des Structures Agraires (FASASA) is central to these schemes, which are administered by the agricultural service in each *département* and to which applications by individual farmers must be made. Agricultural advisors, union officials, solicitors and local politicians are vital in diffusing knowledge of these opportunities.

Special pension supplements (*Indemnités Viagères de Départ*) have been available for more than two decades to encourage elderly farmers to sell their land (Naylor 1982, 1985). The system peaked in 1969 when 80 000 supplements were allocated, involving the sale of 1 200 000 ha, but during the past ten years only about 20 000 additional supplements have been granted annually. Three main factors have contributed to this decline: disruption to family life between 1914 and 1918 produced a small cohort of births – a veritable 'hollow generation' – which reached retirement age recently; inflation has reduced the real value of the supplements; and economic crisis has enhanced perceptions of the security to be derived from working the land. In 1973 special payments (*Dotations d'Installation des Jeunes Agriculteurs*) were introduced to encourage farmers under 35 years of age to acquire holdings of their own. Initially these subsidies only applied to the uplands, but in 1976 a graded system was introduced to cover the whole country, with peak rates in mountain areas, median ones in their periphery, and minimum rates in the lowlands. Additional assistance to favour the installation of young farmers was awarded by the Socialist administration in 1981.

A complementary measure to promote farm enlargement and rejuvenate the agricultural population involves the Sociétés d'Aménagement Foncier et d'Etablissement Rural (SAFERs). They were created following the legislation of 1960 and operate as semi-public, non-profit-making bodies that cover 31 groups of *départements*, embracing the whole of France except Alpes-Maritimes. Their objectives are to regulate the land market and to purchase land as it becomes available in order to enlarge neighbouring holdings or facilitate acquisition by young farmers. They can pre-empt other purchasers and may 'bank' property for up to five years to enable it to be deep-ploughed, fertilized or drained or for new buildings and other facilities to be installed. The average purchase has been a mere 9 ha, but about 1 500 000 ha have been acquired by the SAFERs since 1962, representing under 5% of the nation's farmland and less than one-fifth of all land that has come onto the market. Lack of finance has hindered their operations nationwide, but they have had considerable impact in central and southern *départements*, which have many small, marginal farms. The SAFERs are required to create family farms, which are suitable to employ two people on a full-time basis, and to put these holdings up for sale. Generous loans and subsidies have assisted many young farmers to acquire property of their own, but the SAFERs are unable to help those who would wish to rent land. Since 1970 attempts have been made to operate *remembrement* and other structural reforms in a more harmonized way by arranging a score or so of neighbouring *communes* into *Opérations Groupées d'Aménagement Foncier*.

These fundamental and essentially nationwide reforms have had important implications not just for farming but for rural management in general. *Remembrement* has transformed the landscape of many areas and has raised warnings about the ecological vulnerability of the countryside. The structural measures operating since 1960 have had four main results (Brunet 1974, 1984). They have assisted a modest rejuvenation of the agricultural population, with the proportion of farmers aged over 65 declining from 19.5% (1960) to 16.0% (1984). More importantly, they have accelerated the decline in farm numbers from 1 774 000 to 1 059 000 and have trimmed both the size of the agricultural workforce (from 3 791 000 to 1 659 000) and its rôle in the rapidly changing rural population as a whole. According to the 1982 census, only 7.9% of the labour force worked on the land and 26.5% of the total population lived in administrative areas that were defined officially as 'rural' (i.e. having fewer than 2000 inhabitants). Counterurbanization renders this formal definition rather unhelpful. Various other approaches have been used to define urban areas and to subdivide rural France in the early 1980s (see Ogden 1985). More generous functional definitions of urban areas suggest that only one-fifth of the national population now lives in the countryside.

Area-based plans

In order to complement nationwide policies a number of area-specific plans have been introduced since the mid-1950s. Some serve to intensify the spatial impact of general policies; others involve special organizations for agricultural development, environmental conservation or rural recreation within their operational areas. Despite its title, the *Loi d'Orientation Agricole* (1960) dealt with rural problems which extended beyond the confines of farming. In the spirit of that legislation detailed investigations were undertaken in nine pilot areas and proposals were advanced to enhance farm incomes, introduce new job opportunities (through crafts and tourism), improve rural housing and support service provision in depopulating areas. Collaboration between various administrations would be essential and rural social workers would have a vital facilitating rôle to play (Di Borgo 1966). A handful of 'rural action zones' were defined in Brittany and the Massif Central, and in 1961 mountain districts were recognized as needing special assistance, very much later than in neighbouring Alpine states.

In 1967 this area-based approach was extended to four 'rural renovation zones' (Brittany, Auvergne, Limousin and a scatter of mountainous areas) to which extra development funds were allocated. The four zones covered a quarter of France, contained one-third of its farms and housed 6 500 000 people. Each zone had a special commissioner charged to undertake research and liaise with appropriate ministries and agencies to promote rural well-being. In the early years Brittany was the main beneficiary, with funds directed to new roads, an automated telephone system, accelerated *remembrement*, rural industrialization and tourism schemes. By 1973 greater attention was being devoted to the Massif Central; indeed, two years later a special development plan for the region was announced (Clout 1984). This was partly because certain national figures (e.g. Pompidou, Giscard d'Estaing, Chirac) had their political roots in the Massif, but it was also in response to a new programme for supporting a wide range of rural projects in mountainous districts, which together covered one-fifth of France. The government emphasized that its policy was not based on economic arguments alone but rather on the political imperative of maintaining population in remote areas and the ecological desirability of conserving upland environments. Similar logic had brought special subsidies to French mountain farmers in 1972 and 1974 and was to be reiterated in Directive 75/268 (1975) of the Common Agricultural Policy which authorized financial compensation for farmers in mountains, hilly terrain and other less favoured areas in order 'to ensure the continuation of farming . . . maintain a minimum level of population . . . or conserve the countryside' (Smith 1985) (Fig. 5.1c). During the next ten years several committees of inquiry investigated the human and environmental problems of France's mountain districts, where many area-based programmes were already in operation, culminating in 1985 in the *Loi Relative au Développement et à la Protection de la Montagne*.

The challenge of tackling rural problems on an integrated and rather technological basis has been faced by seven mixed-economy corporations which combine funds from local *communes, départements* and the state with many other forms of finance (Fig. 5.1b). Their enabling legislation dates from the early 1950s, but the Compagnie Nationale du Rhône (1933) provided an earlier model. Its objectives of electricity generation, flood control and farmland irrigation set the technological stance that later *grands aménagements* were to adopt. Their specific objectives varied, but all sought to modernize and diversify the essentially rural economy of their operational areas. Several corporations have encountered serious problems and the most successful have only partially achieved their aims (Tuppen 1983, Laborie 1985). The best known example is the Compagnie Nationale d'Aménagement du Bas-Rhône-Languedoc (1955), whose prime function involves channelling water from the Rhône to irrigate the plain of Languedoc and enable fruit and vegetables to be grown in place of vines. After three decades of activity, the list of achievements includes widespread irrigation, improved farm structures, new crops, modernized marketing and new food-processing industries. But all has not gone according to plan. Only half the projected area has been irrigated and water supplies have proved costly. The new crops are in serious overproduction in the EEC, and the company is supplying more of its water to urban and industrial users than was intended. Some is actually used to irrigate vineyards – inconceivable 30 years ago.

In similar fashion the achievements of the Société du Canal de Provence (1957) have fallen short of expectations. Flooding in the Durance valley has certainly been controlled, but Provençal farmers have been reluctant to invest in irrigated cropping and hence a great deal of water is sent to satisfy urban needs. The Société pour la Mise en Valeur Agricole de la Corse (1957) sought to eradicate mosquitoes on the eastern plain, clear land, install irrigation, lay out new farms and provide technical advice. Its activities proved to be highly controversial, with few islanders showing any interest and most new farmers being *pieds-noirs* from Algeria. Only half of the 25 000 ha equipped for irrigation actually receives water and incoming farmers have favoured growing vines rather than fruit and vegetables. In 1974 the corporation's activities were extended to promote stock rearing in the mountains. The activities of the Compagnie d'Aménagement des Landes de Gascogne (1958) failed even more dramatically, with repatriate farmers successfully suing the corporation on the grounds that the sandy soil was too infertile and that newly laid out farms in firebreaks within the forest were too small for viable farming to be possible. A replacement company was created in 1972 to promote rural development throughout Aquitaine. The three remaining corporations have been far more modest. The Société d'Aménagement des Friches et Taillis de l'Est (1962) reclaimed heathland, set out new farms and promoted afforestation, but its rôle was subsumed by local SAFERs. Irrigation, intensified cropping and improved marketing characterize the achievements of the Compagnie d'Aménagement des Coteaux de Gascogne (1959), and the Société pour la Mise en Valeur de l'Auvergne-Limousin (1962) has introduced a dispersed programme including irrigation, improvements to summer pastures and woodlands, and small tourism projects. Taken together the *grands aménagements* have been a partial success story. Certainly they helped to accommodate repatriates at a particularly difficult time. But their architects placed too much faith in the power of technology and large investment and failed to anticipate local antipathy or the major agricultural developments that were to occur in France's Mediterranean competitors.

Mounting interest in ecology, concern to provide recreation space for city dwellers, and the realization that France was lagging behind her neighbours with respect to environmental conservation gave rise to legislation enabling national parks (1960) and regional nature parks (1967, 1975) to be created. Despite their similarity of title, these area-specific formulae differ substantially in terms of objectives, management and funding (Aitchison 1984). National parks owe their existence to central decisions taken by the Conseil d'Etat and involve areas where conservation of fauna, flora and the natural environment is deemed to be of great scientific interest. Six national parks were duly established between 1963 and 1970 (Fig. 5.4), covering 12 500 km², 2.3% of the whole country. Each has an

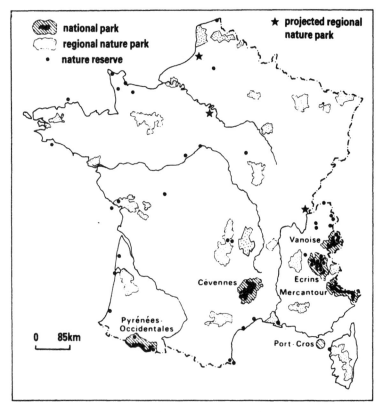

Figure 5.4 National parks and regional nature parks.

administrative committee comprising representatives of ministries, local government officers and members of conservation and amenity organizations. Most parks have a rigorous three-fold zoning structure – an inner protected zone (the 'park proper'), an outer 'envelope', and a number of nature reserves to which only scientists are allowed access. The administrative committee is empowered to manage the 'park proper' but also consults with local authorities regarding the outer 'envelope'. Four parks are located at high altitudes in the Alps and Pyrenees and have virtually no resident population or farming activity in their cores. Much land is owned by the state or local *communes*. By contrast, the Cévennes park contains over 100 farms and a resident population of 600, and Port-Cros park includes several Mediterranean islands and adjacent water.

Protection of flora, fauna and landscape forms the prime objective in each inner zone, which is managed by special staff in receipt of finance from central governemnt. Hunting is banned (with a few exceptions) and the construction of roads, buildings and tourist facilities is forbidden. Less restrictive regulations have been required in the Cévennes where imaginative attitudes by park officers have enabled an effective compromise to be achieved between nature conservation, upland agriculture, craft activities and rural tourism. Rigorous management has been effective in all six inner zones but has not always been popular with local residents (Leynaud 1985). The running of each peripheral zone is entrusted to a committee from the local *département(s)* which seeks to cope with visitors as well as conserving existing landscapes and ways of life. These 'envelopes' are equipped with access roads, information points, visitor services and other activities which have contributed to a reversal of protracted local depopulation since 1975. Visitor pressures are strong and there are powerful demands from new development. Some would argue that generous provision of recreation facilities, accommodation and even factories in some 'envelope' localities has damaged the overall concept of the national park in France.

Regional nature parks respond to rather flexible formulae which were revised in 1975 to embrace conservation of the natural and cultural heritage, provision of recreation space and facilities, promotion of countryside education, and support for rural services and employment opportunities (Daudé *et al.* 1976). Each proposal to create a regional nature park originates with the constituent local authorities and is transmitted upward for approval by regional and, ultimately, central authorities. A charter is drawn up for each park which specifies its perimeter, determines its management body (usually a committee of local inhabitants and users), sets out a management plan and indicates its sources of finance. Most money comes from the appropriate *région(s)*, *départements* and nearby towns, with a modest amount deriving from central sources. A score of parks was set up after 1968 involving 1260 *communes* and covering a total of 28 000

km² (5.2% of France). Each is the product of a unique set of local circumstances and political pressures, with some being equipped for fairly
intensive recreation and others being rather similar to the 'envelopes' of
national parks (Calmès 1978). Many imaginative projects have been
established, including reception and information facilities, *gîtes ruraux*, craft
workshops and facilities for marketing local projects. An impressive range
of 'ecomuseums' has been opened to display vernacular architecture, ves-
tiges of traditional rural activities, implements, furniture and costume.
Grass-roots decision making is extolled as the great virtue of the regional
nature parks, but relative lack of finance is a weakness. Environmental conservation has proved difficult in some localities since management bodies
do not possess the kind of regulatory power encountered in national parks.
It is true that some nature reserves have been established in regional nature
parks, and numerous guidelines for spatial management have been drawn
up. But it must be recognized that new construction and other forms of
rural development are permitted – indeed encouraged in some localities –
provided local guidelines on style and building materials are respected. Of
course, local attitudes also change with time and some *communes* have
pulled out of designated parks (e.g. Armorique), and managing parks
located in more than one *région* (e.g. Marais Poitevin) proves far from
easy.

Recent schemes for rural revival

In addition to 'top-down' nationwide policies and a range of area-specific
schemes, the state devised a number of strategies to encourage *communes* to
link together for purposes of rural management. Since the mid-1970s such
examples of local solidarity and initiative have been viewed with increasing
favour, with new policies, subsidies and contractual arrangements being
introduced to assist rural development, culminating in the special legislation for mountainous areas in 1985. The notion of comprehensive land-
use planning dates only from the *Loi d'Orientation Foncière* of 1967, which
required the preparation of urban structure plans (*Schémas Directeurs
d'Aménagement des Aires Métropolitaines*) to steer future development and
detailed local plans (*Plans d'Occupation des Sols*) to identify precise sites for
conservation or construction and to specify acceptable intensities of building (Fig. 5.5). 'Pressured' rural *communes* on the fringe of urban areas and
stretches of countryside faced with tourism projects or other major
schemes were also required to produce local plans. By virtue of their
designation of development potential – and profit making on land transactions – these documents have proved highly contentious (Dubois-Maury
1983). In fact, most rural *communes* have not prepared local plans but rely on

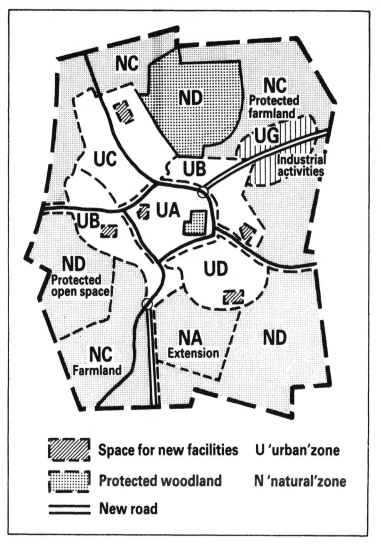

Figure 5.5 An example of a local plan.

simpler and less binding devices such as the designation of a *Zone d'Environnement Protégé* (1976). Legislation enacted under the Socialist régime in 1982 and 1983 reaffirmed the responsibilities of each level of administrative unit and stressed the duty of every *commune* to produce some kind of land-use strategy. These statements enhanced the significance of local plans and the rôle of the local mayor in delivering development permits, while playing down the broad-brush structure plans.

The existence of a very large number of *communes* lies at the heart of French democracy, but almost half have fewer than 300 inhabitants apiece and in remote rural areas local budgets are minuscule. Despite cross-subsidies from richer authorities, few rural *communes* have either the initiative or the finance to set up a development project (Pinchemel 1980). Some *communes* merged during the 1970s, with the total declining from 37 708 to 36 382, but not all were successful and the number of *communes* rose to 36 512 in 1985. By contrast, many rural *communes* have been willing to associate in local syndicates (Limouzin 1980). Most cope with specific issues, such as water supply, waste disposal, electricity supply or bussing of schoolchildren, but some are broad ranging and incorporate guidelines for environmental protection, rural industrialization or new residential development. Their success lent support to the concept of the rural management outline (*Plan d'Aménagement Rural*) which had been advanced in the *Loi d'Orientation Foncière* of 1967. Each outline involves a group of adjacent *communes*, usually 15–20, which are served by one or two market towns. The document is drawn up by local mayors and councillors, members of professional bodies and officers of the Ministry of Agriculture and the *Préfecture*. Each outline reviews current conditions and makes proposals for such matters as infrastructure, services, social housing, building styles, crafts and tourism (Béteille 1975, 1981). By 1980 over 170 outlines had been prepared, involving 6000 *communes* and one-fifth of the whole country (Fig. 5.6). They certainly acted as foci for debate and collaboration, but they were not really plans – in the sense of being legally binding – nor did they incorporate any kind of budget. To put the good ideas they contained into practice, groups of *communes* had to obtain both approval and financial support from external paymasters (Le Coz 1973, 1977, 1984). In 1983 the Socialist administration replaced the management outline with the *Charte Intercommunale*, which also encourages local authorities to work together.

Building on four years' experience of development contracts signed by small towns and the state, the Délégation pour l'Aménagement du Territoire et à l'Action Régionale (DATAR) introduced the notion of the *contrat de pays* in 1975 (Scargill, 1983). These local contracts also facilitated rural co-operation, related to self-defined groups of *communes*, and focused on one or two small towns apiece (Kayser 1979). But in addition, each successful contract incorporated a measure of financial commitment from the state in its central or (since 1978) regional form. By 1982 over 350 contracts had been approved, covering 7500 *communes* in most parts of France and involving a total population in excess of 5 million. Many local contracts had very similar aims to management outlines and essentially replaced them

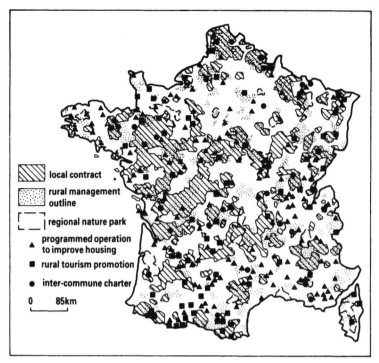

Figure 5.6 Planning collaboration between *communes*.
Source: after Chapuis 1986.

(Vaudois 1980) (Fig. 5.7). For example, detailed objectives of a score of
local contracts in Franche-Comté ranged from environmental conservation, job creation (farming, tourism, crafts, retail) and infrastructure
(piped water, telecommunications) to maintaining public services and providing community facilities for the elderly (Boichard 1985). Yet other
forms of local initiative involve contractual schemes to improve housing
and the delimitation of *pays d'acceuil* within which groups of *communes* co-
operate to promote rural tourism.

Loss of rural services and the desirability of promoting revival in long-
depopulated rural areas were debated many times in French parliamentary
circles during the 1970s and early 1980s. It was stressed that some rural
facilities – such as water supply, electricity and garbage collection – had
improved substantially since mid-century; however, over the same period
numerous village services had been lost (Pitié 1969). Closure of local shops,
schools and post offices and loss of public transport were identified as
major factors contributing to continued depopulation in remote areas

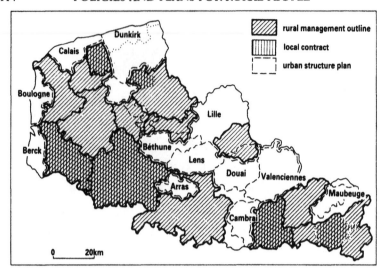

Figure 5.7   Rural management outlines, local contracts and urban structure plans in the Nord.
Source: after Vaudois 1980.

(Chapuis & Brossard 1986). For example, passenger services had run on 37 000 km of railway in 1950 but only 26 000 in 1970, and the proportion of rural *communes* with a post office had declined from a half to one quarter. In 1974 the government instructed each *préfet* to take action to stop the withdrawal of local services in his *département*, and commissions of inquiry were established to report on land use, services, tourism and employment opportunities in rural France. The government announced that at least 6 months' warning should be given before any service would be withdrawn, and its likely implications had to be investigated by a special committee in each *département*. Critical thresholds for maintaining post offices, junior schools in villages and secondary schools in small towns were all revised downwards. The pace of closure was slowed and at present 85% of rural *communes* still have a junior school. Following a decree in 1979, strong encouragement has been given to proposals to combine postal services with activities dealing with social security and other forms of administrative support. Over 2000 examples of using official buildings to accommodate a range of services on a part-time basis were reported in 1979–80. Local initiatives for operating minibuses or taxibuses summoned by telephone receive official backing and financial assistance, with the public transport law of 1982 giving groups of *communes* further support to initiate flexible forms of rural transport. Instead of having to accept that services would be withdrawn or greatly reduced, countrydwellers have been encouraged through the *Opérations-villages* programme to suggest solutions that would meet their needs and are often given practical and monetary support to implement their ideas. In 1979 the Ministry of Agriculture and the DATAR agreed to create the Fonds Interministeriel de Développement et d'Aménagement Rural (FIDAR) to co-ordinate existing subsidies and direct them to rural schemes which respond to local needs in coherent and sustainable ways (Aitchison & Bontron 1984). Thus the FIDAR helps support village stores, workshops and many kinds of service throughout *la France fragile*. Since 1982 the 22 *régions* have helped to finance and steer its activities.

Despite all these measures, many mountainous districts continued to lose population during the 1970s. In addition, there was growing concern that some fragile upland environments were coming under heavy pressure for the installation of winter sports facilities. Certainly the *Plan Neige* of 1970 had opened the way for large new projects in relatively accessible areas with wide expanses of skiable snow, and by the middle of the decade tourist facilities were also being built in remote areas (Grolleau & Ramus 1986). In 1977 the vulnerability of *la montagne* was stressed in an influential speech by Giscard d'Estaing, who argued the case for assisting mountain farmers as logical guardians of the landscape, restricting new developments to sites close to existing settlements, minimizing construction of new roads in the mountains and adopting a more ecologically sensitive approach overall (Barruet 1984, Guerin 1984). Subsequent policies were tailored to the particular circumstances of each massif. Further commissions of inquiry investigated conditions throughout the mountain domain, which covers one-fifth of France but houses only 7% of its population (Besson 1983). However, each year winter sports and summer holidays bring 2 500 000 visitors to join the 3 630 000 permanent residents. The mountains continued to receive attention under the Socialist régime with the innovative *Loi Relative au Développement et à la Protection de la Montagne* being passed in 1985 (Knafou 1985). This stressed the distinctive characteristics and problems of each of France's major massifs (N. Alps, S. Alps, Corsica, Massif Central, Jura, Pyrenees, Vosges) and established a renovation committee for each of them, overseen by a national council. In the spirit of decentralization it encouraged groups of *communes* to liaise with official agencies to produce projects for supporting and creating employment, rehabilitating abandoned dwellings (thereby providing social housing for young couples), and protecting vulnerable environments. Such grassroots ideas are to be articulated through local contracts and assisted with state finance from the Fonds d'Intervention pour l'Auto-développement en Montagne. The new law is important for its holistic stance, but is shot through with questions. For example, how will effective compromises be achieved between economic development and respect for the environment, or between promoting local initiative and enhancing national solidarity?

Prospect

L'aménagement rural has flourished in France over the past three decades, not as a single planning operation but as the manifestation of a host of interlocking policies and plans which have embraced an ever wider range of objectives, agencies and techniques of implementation (Barthélemy & Barthez 1978). With respect to development control and environmental conservation, France was a latecomer among the nations of North-West Europe, but she has made up for lost time and – in the last ten years in particular – has pioneered support for community initiatives and rural services. Equally wide-ranging action will be vital in the future, but it may be argued that two themes will require special attention in the years up to 2000. The first involves the need for effective planning of dispersed suburbanization, which continues to flood into accessible and attractive countrysides. The responsibility for coping with this tide rests firmly with the administration of each individual *commune*. Much vigilance will be required to avoid an almost random scatter of new housing wherever sellers of land, builders of property and enthusiastic purchasers may happen to agree (Berger 1980). Great attention will also need to be paid to the design and construction of new homes. In a far-sighted way, official architectural agencies have been created to advise rural *communes* about the most appropriate kinds of development. The second major challenge involves the necessity to reorientate rural services and facilities toward meeting the needs of the elderly. Rural deprivation is now less a matter of impoverished farming families and is more to do with the presence of so many elderly people in *la France fragile* (Paillat & Parent 1980, Maclouf 1986). Those with local family ties may be able to call on an informal network of support, but retired people who have migrated from distant cities are often exposed to particularly extreme isolation. The need for effective and sensitive social planning remains as great as ever in the French countryside.

References

Aitchison, J.W. 1984. The national and regional parks of France. *Landscape Research* **9**, 2–9.
Aitchison, J.W. & J.C. Bontron 1984. Les zones rurales fragiles en France. *Bulletin de la Société Neuchâteloise de Géographie* **28**, 23–53.

Anon. 1983. L'amélioration des conditions de vie en milieu rural et la protection de la nature. *Journal Officiel* **17**, 1-60.

Barre, A. & J. Vaudois 1980. Autoroutes et agriculture dans la région Nord-Pas de Calais. *Hommes et Terres du Nord* 46-53.

Barruet, J. 1984. La politique de la montagne. *Revue de Géographie Alpine* **72**, 329-46.

Barthélemy, D. & A. Barthez 1978. Propriété foncière, exploitation agricole et aménagement de l'espace rural. *Economie Rurale* **126**, 6-16.

Berger, A. 1975. *La nouvelle économie de l'espace rural*. Paris: Cujas.

Berger, M. 1980. Rurbanisation et analyse des espaces ruraux péri-urbains. *L'Espace Géographique* **9**, 303-13.

Besson, L. 1983. *Politique de développement et de protection des zones de montagne: rapport au Premier Ministre*. Paris: Documentation Française.

Béteille, R. 1975. Un P.A.R. pour les brandes du Montmorillonnais. *Norois* **22**, 116-22.

Béteille, R. 1981. *La France du vide*. Paris: Litec.

Biancarelli, J. 1978. *Aménager les campagnes*. Paris: Le Moniteur.

Boichard, J. 1985. *La Franche-Comté*. Paris: Presses Universitaires de France.

Brunet, P. 1974. L'évolution récente des paysages ruraux français. *Geographia Polonica* **29**, 13-30.

Brunet, P. 1984. *Carte des mutations de l'espace rural français 1950-1980*. Caen: Université de Caen.

Calmès, R. 1978. *L'espace rural français*, Paris: Masson.

Chapuis, R. & T. Brossard 1986. *Les ruraux français*. Paris: Masson.

Chavanes, G. 1975. L'industrie en milieu rural. *Etudes de Politique Industrielle, Documentation Française* **2**, 1-212.

Chevalier, M. 1981. Les phénomènes néo-ruraux. *L'Espace Géographique* **10**, 33-47.

Clout, H. 1984. *A rural policy for the E.E.C.?* London: Methuen.

Daudé, G. et al. 1976. Les parcs naturels français. *Revue de Géographie de Lyon* **51**, 99-204.

De Farcy, H. 1975. *L'espace rural*. Paris: Presses Universitaires de France.

Deniel, J. 1965. Les talus et l'aménagement de l'espace rural. *Penn. Ar. Bed.* **12**, 41-54.

Di Borgo, C.P. 1966. A French experiment in rural development. In *People in the countryside*, J. Higgs (ed.), 109-32. London: National Council of Voluntary Organizations.

Dubois-Maury, J. 1983. Le transfert de coefficient d'occupation des sols. *Norois* **30**, 393-401.

Duboscq, P. & Mathieu, N. 1985. *Voyage en France par les pays de faible densité*. Paris: Centre National de la Recherche Scientifique.

Flatrès, P. 1979. L'évolution des bocages: la région de Bretagne. *Norois* **26**, 303-20.

Fruit, J.P. 1985. Une typologie des campagnes françaises selon la structure de leur population. *Cahiers Géographiques de Rouen* **23**, 73-92.

Grolleau, H. & A. Ramus 1986. *Espace rural, espace touristique*. Paris: Documentation Française.

Guerin, J.P. 1984. Finalité et génèse de la politique de la montagne en France. *Revue de Géographie Alpine* **72**, 323–27.

Hervieu, B. 1978. L'industrie au village ou la banlieue aux champs? *Pour* **60**, 43–9.

House, J.W. 1978. *France: an applied geography*. London: Methuen.

Houssel, J.P. 1978. Aménagement officiel et devenir du milieu rural en France. *Revue de Géographie de Lyon* **53**, 283–93.

Kayser, B. (ed.) 1979. *Petites villes et pays dans l'aménagement rural*. Paris: Centre National de la Recherche Scientifique.

Knafou, R. 1985. L'évolution de la politique de la montagne en France. *L'Information Géographique* **49**, 53–62.

Laborie, J.P. 1985. *La politique française d'aménagement du territoire de 1950 à 1985*. Paris: Documentation Française.

Le Coz, J. 1973. Niveaux de structuration de l'espace rural français: les plans d'aménagement rural. *Bulletin de la Société Languedocienne de Géographie* **7**, 135–67.

Le Coz, J. 1977. Régime capitaliste et aménagement du milieu rural. *Bulletin de la Société Languedocienne de Géographie* **11**, 3–44.

Le Coz, J. 1984. Niveaux de décision et d'organisation dans l'espace rural français. *Cahiers de Fontenay* **35**, 41–52.

Leynaud, E. 1985. Les parcs nationaux: territoire des autres. *L'Espace Géographique* **14**, 127–38.

Limouzin, P. 1980. Les facteurs de dynamisme des communes rurales françaises. *Annales de Géographie* **89**, 549–87.

Maclouf, P. (ed.) 1986. *La pauvreté dans le monde rural*. Paris: Harmattan.

Naylor, E.L. 1982. Retirement policy in French agriculture. *Journal of Agricultural Economics* **33**, 25–36.

Naylor, E.L. 1985. Socio-structural policy in French agriculture. *O'Dell Memorial Monograph* (Department of Geography, University of Aberdeen) **18**, 1–180.

Ogden, P.E. 1985. Counterurbanization in France: the 1982 population census. *Geography* **70**, 24–35.

Paillat, P. & A. Parent 1980. *Le vieillissement de la campagne française*. Paris: Institut National d'Etudes Démographiques.

Pinchemel, P. 1980. *La France* (2 vols). Paris: Armand Colin.

Pitié, J. 1969. Pour une géographie de l'inconfort des maisons rurales. *Norois* **16**, 147–76.

Pitte, J.R. 1983. *Histoire du paysage français* (2 vols). Paris: Tallandier.

Rhun, P. 1977. Destruction d'un paysage. *Hérodote* **7**, 52–70.

Roudié, P. 1983. *La France: agriculture, forêt, pêche*. Paris: Sirey.

Scargill, D.I. 1983. The *ville moyenne*. In *The Expanding City*, J. Patten (ed.), 319–53, London: Academic Press.

Schéma Général d'Aménagement de la France 1981. La France rurale: images et perspectives. *Travaux et Recherches de Prospective, Documentation Française* **81**, 1–164.

Smith, M. 1985. *Agriculture and nature, conservation in conflict – the less favoured areas of France and the United Kingdom*. Langholm: Arkleton Trust.

Tuppen, J. 1983. *The economic geography of France*. London: Croom Helm.

Vaudois, J. 1980. L'aménagement rural dans la région du Nord-Pas de Calais: les plans d'aménagement rural. *Hommes et Terres du Nord* 34–45.

Vergneau, G. 1979. Incidences du remembrement agraire sur l'évolution économique et humaine du milieu agricole. *Bulletin de l'Association de Géographes Français* **460**, 157–62.

6 *The USSR*

JUDITH PALLOT

The historical context of rural policy

The most important event to affect the character of rural life in the USSR after the 1917 revolution was the collectivization of agriculture which took place between 1929 and 1933. During this period millions of independent peasant farmers were forced to join large-scale farm units, *kolkhozi*, bringing with them their land, livestock and capital. Henceforth, they had to work together in the fields, selling their produce to state purchasing organizations and distributing the income made among themselves. Collectivization transformed rural society and rural politics, replacing rich peasants, priests and village elders by a new generation of leaders, represented by Communist party cadres, collective farm managers and 'brigadiers'. Villages, hamlets and dispersed farmsteads were incorporated into the territory of collective farms, and in the east nomads were forced to adopt a sedentary life. The economic purpose of these transformations was to reorganize agriculture in such a way as to provide for the transfer of resources from the countryside to the towns. For the next few decades agriculture was 'squeezed' for the sake of heavy industry. Rural living standards suffered accordingly; agricultural workers became second-class citizens denied many of the legal rights of town dwellers, and they were poorly remunerated for their work in the collectives. In more recent decades agriculture has received large injections of capital, and wage reforms have meant that the standard of living of the rural population has been able to rise. Other reforms have removed many of the restrictions that formerly existed on the mobility of the rural population. Yet rural living standards remain relatively low in the pecking order for the allocation of resources in the USSR, and this continues to put limits on the realization of rural policies.

The first collective farms were usually formed from existing village communities and their surrounding land. In the years that followed, farms grew in size as a result of mergers, or they were converted into state farms, *sovkhozi*. These are much larger units which, by virtue of their greater degree of incorporation into the economic planning system, have always been considered more socialized than collectives. Currently, there are 22.9 thousand state farms in the USSR, with an average size of 35.2 thousand hectares and 26.7 thousand collective farms with an average size of 9.1 thousand hectares (*Gosudarstvennyi Komitet SSSR po statistike* 1987: 208, 222). The growth of farm size since the 1930s has meant that the number of separate settlements found within a farm's territory has increased. Today, individual collective and state farms can have twenty or more villages and hamlets on their territory (Konchukov 1979: 28), and in the Baltic states and north-west, where rural settlement has traditionally been dispersed, farms can contain a multitude of scattered dwellings. Only 15% of farms have just one settlement apiece (Kovalev & Koval'skaya 1980: 226). Normally there is a central settlement on every farm in which the administration is located and to which other settlements are subordinated. These latter can consist of section villages (*poselki otdelenii*) and brigade villages (*poselki brigad*), from which the farm's subdivisions are administered and worked. There can also be a scattering of other settlements such as seasonal work stations and mobile dormitories, as well as hamlets and individual dwellings. The various settlements found on the territory of collective and state farms develop, 'according to a general plan . . . each settlement constituting a different link in one large agricultural enterprise' (Kovalev & Koval'skaya 1980: 226). Each such territorial system is subordinate to a district centre, often a small town, in which branches of central planning agencies and economic ministries are located. The centre is responsible for the administration of the farms in the district, and it is usually the site of food processing industries, the centre of transportation and local trade and the focus of the district's cultural and social activities.

It would be a mistake to assume that everyone living on the territory of *kolkhozi* and *sovkhozi* is a farm employee or that all rural settlements in the USSR, below the level of district centre, are farm villages. This may have been the case in the 1930s, but changes since then have complicated the picture. Firstly, the farm labour force has become more diversified since the 1930s as financial specialists, agronomists and engineers have swelled the ranks of the manual workers. Non-agricultural personnel, who are not necessarily farm employees, have also arrived in the villages; they work in education, retailing, health and other services. Secondly, non-agricultural settlements have emerged which cater for recreation and leisure. Thus, in the suburban zones of large cities second-home settlements have grown up and in scenically attractive areas, sanatoria and tourist settlements. A third change has come in the past two decades as the government has encouraged groups of farms to join together to develop rural industries. This has resulted in the emergence of agro-industrial villages (*agrarno-promyshlennie poselki*). Despite these changes, the majority of rural settlements in the USSR are still located within the boundaries of collective and state farms, and the lives of their inhabitants, whether or not they are farm employees, are caught up in the economy and power structure of the Soviet agricultural system. As if to recognize this fact, Soviet law categorizes second-home settlements, mining and resort villages as 'settlements of urban type' (*poselki gorodskogo tipa*) rather than as rural settlements (*sel'skie poselki*). The former are covered by the same land-use regulations as towns, and these regulations differ from those covering rural settlements (Erofeev 1970: 22). Such 'settlements of urban type' numbered 3863 in 1981 as compared with many tens of thousands of mainly agricultural settlements distributed among 22 710 rural districts (*Naselenie SSR* 1983: 22). The total population of these rural districts numbered 96.3 million in January 1985, making up 34.5% of the Soviet population (*Tsentral'noe Statisticheskoe Upravlenie SSSR* 1985: 5), of which up to two-thirds were directly or indirectly involved in agriculture.[1]

Restructuring the rural settlement network

Nikita Khrushchev was the first Soviet leader to give serious consideration to the quality of life in rural USSR. By the middle of the 1950s, when he came to power, there were compelling reasons why rural policy should be placed on the political agenda, for in the postwar years a deep crisis had developed in agriculture which was threatening to undermine the whole economy. One aspect of agriculture's crisis was the shortage of labour. Industrialization had required the transfer of people from rural to urban areas, and between 1927 and 1958 61 500 000 peasants left their villages to join the urban labour force. In subsequent years the flow has lessened, but still 37 million people left for the towns in 1959–78, the most recent intercensal period (*Naselenie SSSR* 1983: 31). By selecting out the young and able-bodied, migration has left many farms in the Soviet Union with seasonal shortages of labour, which have to be relieved by drafting soldiers and students to work on the land at harvest time, and undersupplied with mechanics and other skilled workers. Efforts to encourage reverse migration have so far failed to make an impression on the negative migration balance in rural areas. In Khrushchev's time the diagnosis of the problem was the differential living standards between rural and urban areas. Although the worst discrimination against rural inhabitants is over, the poor quality of rural life continues to be given by Soviet experts as the reason for rural out-migration, which has been exacerbated by rising consumer expectations in the USSR. The author of one newspaper article noted that though two decades previously poor roads were considered 'an organic part' of rural life, in the 1980s they were no longer tolerable (*CDSP* 1984a: 14). The first priority of rural policy in the past three decades has, therefore, been to raise the standard of living in rural areas with a view to stabilizing the agricultural labour force. Concerns about recreation, conservation and second-home development also figure in rural policy but, in a country which still experiences agricultural problems, these issues remain secondary to the principal concern about the quality and size of the rural labour force.

The attempts of Soviet planners to raise living standards have, for the past 30 years, been linked to a programme of village concentration. The programme was launched at the end of the 1950s and sought to eliminate small villages, hamlets and isolated dwellings from the countryside. Under it the settlements in a rural district were classified as either 'viable' (*perspectivnie*) or 'non-viable' (*neperspectivnie*). The former were settlements for which future expansion was planned and which were to receive improved rural services, and the latter were settlements for which no future was envisaged; they were to be left to die out and their populations encouraged to move to the viable villages. Both economic and ideological arguments lay behind the village concentration policy. The economic argument was based on cost comparisons for the provision of services to villages which showed that costs began to rise sharply when village populations fell below 1000 (Pallot 1979: 217). Since at that time over half the USSR's rural settlements had under 100 inhabitants, to reap the economies of scale in service provision called for the mass resettlement of population in a vastly reduced number of rural settlements. Khrushchev was not deterred by the prospect of such a radical restructuring of the rural network; indeed, his earliest proposal on rural policy was to rehouse the rural population in 'agrotowns' of 10 000 or more population. Soviet ideology, which since the 1930s has asserted the superiority of the town as a form of settlement (Pallot & Shaw 1981: 240), seemed to legitimize any measures that would move villages qualitatively nearer to towns. As the 1961 Communist party programme put it:

> Collective farm villages must be gradually transformed into merged populated places of the urban type with good housing, services, cultural and medical institutions. The rural population must ultimately be brought up to urban standards in terms of cultural and service aspects of living conditions. (*Programma KPSS* 1961: 85)

At the time that village concentration policy was approved there were no formal procedures for physical planning at the local level. Most building, whether of housing, production facilities or service buildings, on collective and state farms was undertaken on an *ad hoc* basis by the farm authorities. Similarly, off-farm construction by local authorities and economic ministries took place without reference to an overall plan. In 1959 this situation was changed and, thereafter, every rural district had to be furnished with a long-term plan for land use and the provision of services (*raionii plan*). These plans, which have been drawn up for all rural districts in the USSR, are the work of special planning commissions made up of representatives of the local authorities, architects, agronomists and farm managers (Konchukov 1979: 17). They cover the distribution of the settlement network and of cultural, health and service establishments, the location of rural industries, the development of the road network, the water system and electricity system and the provision of open space. Although drawn up for 15 to 20 years ahead since 1959, district plans have been subject to frequent modifications.

The fulcrum of the first generation of district plans was the provision they made for the evolution of the settlement system; projections for village growth or extinction were needed before health, education, cultural and other services, roads and water supplies could be planned. It was in the settlement plan that villages were classified as either viable or non-viable. The classification effectively prohibited capital construction in non-viable villages and required a land-use plan to be drawn up for each viable village. The choice of category for individual villages was determined by two factors: their size and function. The largest, or those housing the farm administration, were usually designated as viable. Small villages, hamlets and individual dwellings were typically classified as non-viable and, in some plans, brigade and section villages were included among the non-viable as well. One author has estimated that 85% of all collective and state farm settlements were classified as non-viable as a result of the first-round of district planning (Balezin 1972: 599). The figure was clearly too high; to concentrate the Soviet rural population in 15% of its existing rural settlements was unrealistic. As soon became apparent, the economic costs involved in uprooting and relocating millions of rural inhabitants outweighed the benefits; increases in the lengths of journeys to work caused problems for farms, and there were obvious social costs (Pallot 1979: 218–20). Furthermore, it transpired that many district plans had been hurriedly put together and were based on inadequate social surveys and demographic analyses. Such criticisms prompted the post-Khrushchev leadership, in 1968, to order a review of district plans. The outcome was that the number of villages designated viable was increased, according to one authority, by one half. As of 1975, the figures for viable villages in a selection of Soviet republics was as follows: Russian Republic 30.4%, Ukraine 26.2%, Belorussia 18.6%, Estonia 11.4%, Latvia 0.85%, Turkmenistan 37.5%, Azerbadjan 34.9%, Georgia 85.1%, Moldavia 64.2% (Rogozhin 1972: 10). The variations reflect differences in the existing degree of settlement dispersal between Soviet republics, as shown in Figure 6.1. The commitment to the idea of village concentration remained and was given a fillip in 1975 by an ambitious programme for the integrated development of the non-black earth belt in European Russia. The proposals for rural settlement in the programme followed the established lines. Thus, the rural population, scattered at that time in 140 000 small rural settlements, was to be resettled in 17 500 villages of 500–2500 in size. In Kalinin region, as an example, the existing 12 000 settlements were scheduled to be reduced to 1000, involving the resettlement of 8500 families in the first five years (Tobilevich 1979: 49).

It is difficult to learn from Soviet sources the extent to which village concentration plans have been realized, or their economic and human consequences. It is unlikely that a resettlement programme on such a scale could be carried out without opposition. Hints of the disruption caused by the policy have begun to emerge in the past few years. Tobilevich, a specialist in village planning, wrote in 1979 that the resettlement of people in the non-black-earth programme 'can be compared with collectivization in terms of its socio-economic consequences' (Tobilevich 1979: 48). Describing one plan in Siberia, which involved the resettlement of the population of 65 villages, Tobilevich admitted that the inhabitants forced to leave their homes 'did not comprehend the utility or necessity of change'

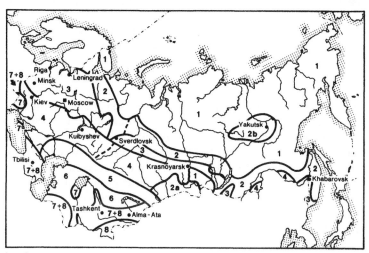

Legend
1   Sparsely settled forested region with mining and herding villages
2   Agricultural and wood-processing villages of the settled periphery of the Russian *taiga*
3   Non-black-earth agricultural region with many villages; c.75% have less than 100
    inhabitants and there are many isolated dwellings
4   Densely settled agricultural region of the black-earth belt. More than 50% of the villages
    have more than 200 inhabitants and 20–40% have more than 500. The eastern section
    includes the Virgin Lands
5   Dry steppe and semi-desert pastoral farming zone with sparse small settlements
6   Desert pastoral farming zone with sparse small settlements and larger villages in oases
7   Southern zone of intensive irrigated agriculture with dense population and large villages,
    many greater than 1000 in size
8   Mountain-farming zone of sparse settlement

Figure 6.1   Zonal characteristics of rural settlement in the USSR.
Source: after Kovalev & Koval'skaya (1980:220).

and that 'administrative measures' had had to be employed (Tobilevich 1979: 54). In some places changes took place rapidly; in Kirov region, for example, the number of villages was reduced from 11 250 to 9480 in the space of three years, a feat achieved by shifting the peasants' old homes on sledges from the non-viable to the viable villages (Tobilevich 1979: 49). Elsewhere lack of funds meant that village concentration never progressed beyond the planning stage, but the force of the policy was nevertheless felt, the mere designation of a place as non-viable branding it as inferior and depriving it of resources for capital construction and even basic maintenance. Tobilevich (1979: 50) drew attention to the tendency for schools, medical services and shops in non-viable villages to be closed long before their population departed and for their transport links to be severed. Such a situation was described in a letter to 'Rural Life' (Sel'skaya Zhizn'), the Soviet Union's daily rural newspaper. In it a machine operator from a collective farm in one region described the demoralization of the population in his native village since it was designated 'non-viable' in 1978:

> What sort of grandiose programme is this that rides roughshod over the fate of so many people? We want to be good workers to look after the land and make it more productive . . . but we are weak compared with the collective farm management. (Sel'skaya Zhizn' 1985c)

Whether it is because of human factors or the economic costs involved, in the 1980s the policy of village concentration has fallen into official disfavour. This was signalled by articles and letters, like the one cited above, that began to appear in 'Rural Life' in 1985. Journalists for the newspaper began to write about the advantages of small villages, recording their 'embarrassment' and 'discomfort' at the thought that such settlements were previously designated non-viable (Sel'skaya Zhizn' 1985b, c, d). In January 1985 the newspaper began a series of profiles of small villages, prefacing the first with the following justification:

> In the 1980s more and more frequently we hear voices defending small enterprises and small settlements. This is because having moved everything into the central farm settlements, the workforce has been distanced from the fields, and with a weakly developed road network, such a concentration of production not infrequently has led to difficulties (Sel'skaya Zhizn' 1985a).

It is interesting to observe that this reservation about village concentration policy had been voiced by S.A. Kovalev (1968a, b), Professor of Geography at Moscow University, over two decades previously.

The type of policy that will replace village concentration must remain an open question at present. Communist party general secretary Mikhail Gorbachev, in his speech to the 27th Party Congress in 1986, reaffirmed the commitment to raising rural living standards, but left unsaid the means by which this was to be achieved. The district plans drawn up in 1959 and revised in 1968 must now largely be defunct. In one district in the Moscow region the intention to revise yet again the district plan has been announced (*Sel'skaya Zhizn'* 1985a), and it can be assumed that other districts will follow suit. Alternatives have been suggested to village concentration that would improve access to services. Khodzhaev & Khorev (1972) of Moscow University have developed the concept of the 'unified settlement system', in which services are distributed among an interconnected set of settlements, and ideas akin to the 'key settlement' approach have been aired in the press (*Sel'skaya Zhizn'* 1985a). The alternative that seems to be being followed in some districts is to attempt to upgrade services in all settlements, however small. Articles now appear lauding 'the six-pupil school' and 'kindergartens for eleven children' (*Sel'skaya Zhizn'* 1985a). In current writing about rural areas in the USSR, all reference is to the introduction, rather than the running down, of services, but to date no indication has been given of the total cost of taking services to all the USSR's rural settlements or of how the costs are to be borne.

Planning the 'socialist village'

During this century, life in the Soviet Union has been punctuated by periods of upheaval that have left villages denuded of their populations or physically destroyed. The revolution and civil war, collectivization and World War II were just such episodes; the German invasion of 1941, for example, resulting in the loss of over 1 million rural homes and tens of thousands of hospitals and schools (Konchukov 1979: 11). In the aftermath of these episodes the need to reconstruct has enabled planners to experiment with village design. Other opportunities have been afforded by the village concentration policy and by the agricultural colonization of virgin lands, such as took place under Khrushchev in the 1950s and 1960s. Reality and ideology have combined to push the USSR towards a 'goal-oriented' approach to the planning of rural settlements. The physical planning of villages, in fact, dates from the 18th century, when Peter-the-Great drew up regulations for rural construction largely for fire prevention. In the decades that followed further regulations covering all elements of the village's physical structure were passed. A mid-19th century plan for a new village in Siberia is shown in Figure 6.2.

Imitating town design, village plans typically consisted of a gridiron street layout, with a central square on which the church and government

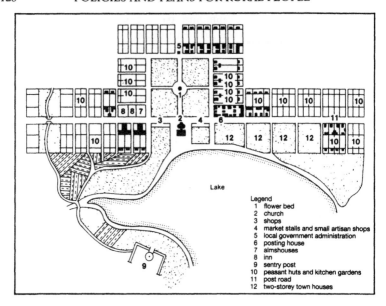

Lake

Legend
1 flower bed
2 church
3 shops
4 market stalls and small artisan shops
5 local government administration
6 posting house
7 almshouses
8 inn
9 sentry post
10 peasant huts and kitchen gardens
11 post road
12 two-storey town houses

Figure 6.2   Plan of a Siberian village in 1830.
Source: after Konchukov (1979: 8).

buildings were located, symbolizing secular and spiritual authority. The
number of villages built according to such plans was small and was confined
to administrative centres (Tobilevich 1979: 39), although all over Russia
attempts were made to enforce fire regulations. The majority of peasant
villages were places which had grown up haphazardly and consisted of
winding streets with mean wooden or wattle-and-daub peasant huts and
adjacent kitchen gardens together with their sheds and barns (Konchukov
1979: 7).

After 1917, village planning entered a new phase as pre-revolutionary
ideas were combined with a new desire to create distinctively socialist
settlements in the USSR. Although most of the architectural community's
efforts went into town design, collectivization saw the publication of plans
for 'model' collective farm villages. Two new elements made their
appearance in villages in the wake of collectivization. The first was the
'production complex', a collection of buildings housing collectivized livestock, farm implements and machinery, stores and workshops. The second
was the 'social centre', intended for the 'cultural and ideological education'
of the rural masses. The introduction of these elements was not at the time
accompanied by wholescale village reconstruction; new farm buildings
were put up wherever land was available, usually on the periphery of

villages, and social centres were accommodated in existing buildings (Tobilevich 1979: 43). Systematic planning dates from the early 1960s and evolved in conjunction with the village concentration programme. Under the programme, all 'viable' villages were required to be furnished with a long-term physical plan (*genplan*) which was to be drawn up by specialist architectural organizations in the Ministry of Construction and approved by the local government authorities. An important component in all village plans is land-use zoning. Characteristically, a division of village territory was made into the 'production zone' (*proizvodstvenniya zona*) and 'settled zone' (*seliteb'naya zona*) separated by a green belt (*zashchito-sanitarnaya zona*). As in the early years of collectivization, the production complex contains farm buildings, including sometimes the farm administration, veterinary and other agricultural support services; the settled zone contains dwellings, social and cultural buildings, retail outlets, sport facilities, health and education services. The aim of physical planning is to achieve certain economic and aesthetic goals. These include cost-efficiency, economy in the use of land, provision of basic services and the accommodation of 'private plot', or allotment, agriculture (Tobilevich 1979: 43). According to one authority, 'the village plan must be compact, it must be based on a precise architectural concept and have a singular composition with the social and cultural centre in a prominent position, an uncomplicated network of streets, alleys and entrances to groups of houses or to individual buildings' (Konchukov 1979: 49). The village centre in particular must 'invoke in all the inhabitants a feeling of pride in their village' (Konchukov 1979: 49). Villages where development conforms to these various principles are termed 'settlements of the new type' (*poselki novogo tipa*) in Soviet literature, and the broad aim of the authorities since physical plans began to be drawn up is to reconstruct all 'viable' villages into 'rural settlements of the new type'. It must be assumed that with the abandonment of the village concentration policy these aims now are being extended to villages hitherto classified as non-viable.

The precise arrangement of land use, types of buildings, the range and level of service provision and design features of 'rural settlements of the new type' has not remained unchanged since the 1960s. The first round of village planning produced plans which were influenced by contemporary ideas in town planning. Designs for single urban micro-regions were used to plan whole villages, street patterns were uniformly geometrical and housing undifferentiated. Further, there was a tendency for the central square, with its social and cultural buildings to be disproportionately large for the surrounding dwellings (Tobilevich 1979: 44). The pioneer settlements that were constructed in the Virgin Lands from the 1950s suffered from these problems. Thus, 500 central settlements were constructed on new state farms in Northern Kazakhstan according to a 'rigid pedantic geometry' (Tobilevich 1979: 188), with long streets intersecting each other at right angles and low-density construction exposing buildings to the dry steppe winds.

In the Virgin Land settlements, dwellings were either bungalows for families or hostels for single people. Elsewhere in the Soviet Union new housing needed for people re-settled from non-viable villages was more typically in the form of blocks of flats which resembled the five-storey pre-fabricated blocks being constructed at that time in towns. The advantage of using blocks of flats from the civil engineering point of view was that they were easy to service and their construction represented a considerable cost saving per unit of floor space compared with separate houses (Tobilevich 1979: 87). Their disadvantage was that they were unpopular with rural inhabitants because they severed people from their 'private plots'. These are the allotments of land (0.25–0.5 has) allocated to agricultural workers and some other types of rural dwellers for their personal use. In Stalin's time private plots constituted a lifeline for collective farm workers, providing them with a means of subsistence; today they provide an important source of supplementary income for farm workers. It is not surprising that any development threatening to make the use of the private plot more difficult should be resisted by the rural population. In the village plans of the 1960s, often no provision was made at all for plots or, if they were included, they were located inconveniently with no storage sheds or livestock shelters. One study made in the 1970s showed that, presented with a choice between a fully appointed flat with an allotment at 150 metres distance and less well appointed separate dwelling with adjacent plot, the preference among a varied group of interviewees was for the second of these options (Spektor & Tomilin 1974: 114–30).

The difficulties experienced in the 1960s indicated the absence of a tradition of village planning in the country. The task of designing 'socialist' villages had been given to architects and engineers who were trained in urban planning but had no experience of rural life. In more recent years some of the early problems have been overcome. According to Soviet authors, today's village plans show greater attention to the special features of the local relief; they take up themes from national traditions of building and design, and they are based on more thorough analyses of the population's preferences than previously (Konchukov 1979: 12). Figures 6.3, 6.4 and 6.5 reproduce plans currently being implemented for three villages in different ecological and cultural settings. The first is of Snov village, the central settlement of a collective farm in Belorussia. Construction began in Snov in 1956. The intention was to develop the village to receive the farm's population of 5000, which at that time was dispersed among 17 small villages and 800 isolated dwellings. In the plan the village axis is an avenue leading to a central park with public buildings. The village is provided with a wide range of services – a sports area with swimming pool and stadium, a school, kindergarten, shops, a social centre and dormitory block – and housing takes the form of two- or three-storey maisonettes and bungalows. Each household's private plot in the Snov plan is divided into two unequal parts, the smaller either abutting onto the house or at a short distance from it and the larger, together with outbuildings, beyond the residential areas. Snov's plan can be said to typify plans for central farm settlements in the forested regions of the USSR. It is here that the village concentration policy involved the most radical construction programme and this, coupled with the premium placed on the land in the region, meant that maximum compactness had to be aimed for in settlement design.

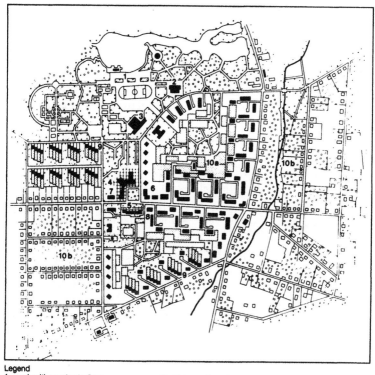

Legend
1 park with sports stadium
2 swimming pool
3 school buildings
4 social and administrative complex
5 architectural monument
6 shops and services
7 dormitory block
8 four-storey, multiple flat blocks for young couples
9 bungalows and two–three-storey maisonettes
10a, 10b private plots

Figure 6.3 Plan of Snov village, Kalinin Collective Farm, Minsk region, Belorussia.
Source: after Tobilevich (1979: 149).

Figure 6.4 is of a village in the entirely different cultural and environmental setting of Uzbekistan in Soviet Central Asia. The central settlement of State Farm No. 21 is located in the open, flat land of the Hungry Steppe. The plan shows the move away from the geometrical street layout that characterized plans of the 1960s for such regions, and it provides park areas to help modify the microclimate. The location of a tea-house (*chaikana*) is a concession to local custom. Housing in the village has been especially adapted to the large size of Central Asian families: two-to-six-room maisonettes and equally large detached and semi-detached houses, each dwelling with its own private plot attached.

At the other end of the country from State Farm No. 21 is the village of Dainava, shown in Figure 6.5, which is situated in the republic of Lithuania in the heart of the dairy-farming region. The plan of Dainava gives the village the same complement of services as the other two villages. The interesting element in the plan for Dainava is the provision made in the village of two blocks for collective farm workers' privately owned livestock. These lie on the edge of the 'settled zone', but are part of it and separate from the collective farm's livestock complex in the production zone. One point of discussion among legal experts in the Soviet Union has

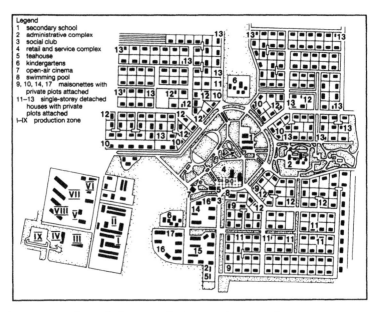

Figure 6.4 Plan of the central farm settlement of State Farm No.21, Uzbekistan.

Source: after Tobilevich (1979: 226).

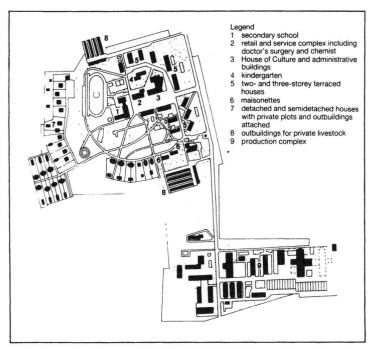

Legend
1 secondary school
2 retail and service complex including doctor's surgery and chemist
3 House of Culture and administrative buildings
4 kindergarten
5 two- and three-storey terraced houses
6 maisonettes
7 detached and semidetached houses with private plots and outbuildings attached
8 outbuildings for private livestock
9 production complex

Figure 6.5 Plan of Dainava, central farm settlement of Leonpolis State Farm, Lithuania (constructed 1965–71).
Source: after Tobilevich (1979: 269).

been whether land used in private-plot farming constitutes part of the state agricultural land fund (Kovalenko 1970: 42). The arrangement in Dainava is indicative of the importance attached at the village level to maintaining the distinction between the socialized and personal sector in farming.

It is clear from the plans shown in Figures 6.3, 6.4 and 6.5 that planners now favour low-rise housing which enables the occupiers to gain direct access to their private plots. In part, the retreat from the earlier tendency to build upwards must be due to popular resistance to high-rise housing, but it also reflects a change in policy towards the private sector in Soviet agriculture from one of 'reluctant tolerance', which characterized the greater part of the Brezhnev period, to one of open encouragement in the past few years. This policy shift is a consequence of the continuing failure of socialized agriculture to produce certain types of agricultural commodities in sufficient quantity and a general move towards encouraging individual initiative in the economy at large. Thus it is that authors writing about rural planning can maintain that collective farm workers' private plots 'are the fundamental element in the settled zones of collective farm settlements' (Konchukov 1979: 67) and that 'for the good of society' it is important to ensure that plans for village reconstruction include provision for families to carry out subsidiary farming (Tobilevich 1979: 82). An article in a leading economic journal recently bemoaned the fact that 106 villagers living in a multi-storey block on one state farm possessed between them only one cow and 50 piglets (CDSP 1983a: 15). Despite its clear attractiveness on cost grounds, flat building on a large scale is coming to an end in rural USSR. Precise figures of the number of rural families that were rehoused in such blocks during the preceding decades do not exist. It is known, however, that at the turn of this decade approximately three-quarters of the Soviet rural population was housed in separate dwellings with adjacent plots which included both old and new houses; the remaining quarter, it must be assumed, were in blocks of flats, dormitories or other multiple occupancy dwellings (Tobilevich 1979: 81).

Rural issues and conflicts

In the discussion above, the division of responsibility for the various phases of village reconstruction has only been hinted at. The principal parties concerned are the farm managements on whose land rural settlements are located, local government organizations and rural residents. The interplay between these parties is difficult to trace with any certainty, but some trends have been discernible for the past 30 years. As has already been observed, one consequence of the development strategy pursued by the Soviet state from the 1930s was the neglect of rural areas and their problems. Although local authorities were given responsibility for many aspects of rural development, decisions about the use of land within the boundaries of collective and state farms were taken by farm managements, under the direction of the Ministry of Agriculture. The use of land under settlements on farm territory, the development of internal farm roads and other amenities, the size and location of private plots and, from the 1960s, the putting together of village plans were under their control (Balezin 1972: 55). Thus, in 1965, the Communist party discussion journal was able to observe that 'the allocation of building land for houses and for cultural and service establishments as a rule is done by the managements of collective farms and the directors of state farms without the participation of the local council's executive committee' (Partinaya Zhizn' 1965: 17). Balezin (1972: 16), a lawyer at Moscow University, made the same observation, noting that farms had been forced to take responsibility not only for rural construction but for the provision of the services as well: 'The fundamental responsibility for providing cultural, social and communal services to the population living in rural settlements rests with the agricultural enterprise.' The popular view of rural settlements was that they 'belonged' to particular collective or state farms, although no such legal category in fact existed.

The introduction of agricultural reforms under the leadership of Brezhnev and Kosygin in the middle of the 1960s resulted in a general review of legal relationships in rural USSR which culminated in the publication in 1968 of a new land law and in an extension of local authority powers under regulations passed the same year. Under the new land law, land lying under rural settlements was made a distinct legal category, separate from the state agricultural and forest land fund, and subject to the control of local authorities. Henceforth, any enterprise or individual wanting an allotment of village land had to apply to the local authority for permission. Local authorities further were to adjudicate in disputes about land and were given powers to order the cessation of illegal building in villages (*Pravovoi Rezhim* Zemel' v SSSR 1984: 162). For all these changes in the legal status of village lands, literature published in the years either side of 1968 tells of the inability of local authorities to take on the tasks assigned to them. One author, acknowledging the poor record of collective farm managements in rural building, nevertheless argued:

> In the existing situation the question of removing (village) land from collective farms is premature; the removal of such land from the control of farms would destroy their interest in village improvement, whilst the state does not have the strength or means to carry out such work at present. (Kovalenko 1970: 120).

Others observed that local authorities did not have the personnel needed to draw up village plans (*Pravovie Problemy Rukovodstva i Upravleniya Sel'skogo Khozyaistva v SSSR* 1970: 213, Balezin 1972: 71). The result was that many villages remained without plans, and building continued to take place without reference to the plan in those that did (Balezin 1972: 77–8). A further limitation on the control local councils could exercise was that non-viable villages were not covered by a land-use plan and decisions about how they were run down effectively rested with the Ministry of Agriculture, operating through farm managements. If the law was deficient in respect of non-viable villages, some lawyers were dissatisfied with the provisions for viable villages as well. The problem was that although village land had been set apart from farm land and forests its different internal uses were not recognized in law beyond the crude division into the 'productive' and 'settled' zones (*Pravovoi Rezhim Zemel' v SSSR* 1984: 162).

This has made the task of resolving land disputes and monitoring building activity difficult and left open the fundamental question of who is responsible for initiating and financing rural construction.

Recent Soviet articles make clear that for the central authorities building in villages should be a joint venture with local councils, farm managements and individual rural families contributing towards the cost and taking an active part in the realization of village plans. Newspapers regularly contain accounts of successfully completed joint projects for the construction of clubs, sports centres, shops and public buildings and, increasingly in recent years, of private housebuilding supported by public and farm funds. These accounts speak of the expansion of local authority involvement in village development compared with the past and of the increasing liquidity of farms and of the rural population. However, in a situation where precise legal responsibility for various elements of village land-use remain ambiguous, difficulties can be expected to arise as the various parties attempt to dominate or to distance themselves from the task of village development. A recurrent complaint in the popular press is the diversion into production of funds earmarked for the purpose of public and service buildings by farm managements. In one article in 'Rural Life' for 1980 two neighbouring villages located on different collective farms were compared; in one the farm management had developed all the services contained in the village plan, but in the other the management, 'had not allocated any resources to the development of services for the population or to civil engineering works' with the result that many young people had left and the school was under threat of closure (Sel'skaya Zhizn' 1980b). Another article reported that farms in one region which had been instructed by the local authorities to supply gravel for road construction had met only half the required quota; 'the managers of the farms' the author complained, 'in effect, ignored the decision of the executive committee of the district council' (Sel'skaya Zhizn' 1980c).

Sometimes it is local authorities that come in for criticism, 'Rural Life' taking up the cause of one village that on repeated occasions asked for their road to be made up because it was impassable during the spring thaw. Despite promises to see to the road's construction, nothing more had been heard by the villagers (Sel'skaya Zhizn' 1980a). A common complaint has been of the failure of local authorities to ensure that village plans are carried out. To be fair, their efforts in this direction are hampered by shortages of funds and their consequent reliance on farm managements to bear the cost of rural construction. Local authorities have encountered particular problems in the control of individual housebuilding. Private house ownership is widespread in rural USSR. Between 1965 and the end of the 1970s, 45% of houses put up in rural districts in the Russian republic were privately owned, compared with 10% in the ownership of collective farms and 45% in the ownership of the state. In the USSR as a whole the equivalent figures were 60%, 6.6% and 33.3% (Tobilevich 1979: 79). Current policy is to encourage further private housebuilding. Measures in 1981 and 1983 were introduced to extend financial assistance to families for private housebuilding and to help towards the cost of linking up with mains water, sewage and gas systems (*CDSP*, 1983c: 23). Larger than average payments were reserved for families moving from non-viable to viable villages and for migrants from urban areas. In the past, much private house building took place without reference to village plans and involved illegal land transactions (Nikishina 1971: 9–10). Thus, for example, private houses have been put up on land designated for other purposes (*SDT* 1964: 71), extensions made without planning permission being sought (Fig. 6.6), and building lots traded in strict violation of Soviet land law, which vests the ownership of all land in the USSR in the state.

Local authorities have sometimes found their decisions overridden by farm managements. One such case involved a schoolteacher who defended his decision to build his house in a place different from the one assigned by the local authority by arguing that the new site was approved by the collective farm (*SDT* 1959: 97–8). This illustrates the popular perceptions current at the time about the final arbiter in village planning. Illegal housebuilding and land conveyance does not necessarily work in favour of the individual. A newspaper article in 1983 contained an account of the misappropriation by a collective farm management of its members' private

Figure 6.6 'Illegal house extension' (cartoon from *SDT* 1964: 17).

plots for the construction of second homes for urban dwellers, apparently with the connivance of the local government authorities (*CDSP* 1983b: 16). A similar report appeared in 'Rural Life' three years later (*Sel'skaya Zhizn'* 1986).

The pressure on land for second-home construction has emerged as a major issue in the past decade as incomes and aspirations have risen among the Soviet Union's urban population. Large towns in the Soviet Union are encircled by 'dacha', or second home, settlements to which urban dwellers move for the summer months in line with old Russian customs. In recent years the land assigned by the state for the development of such settlements has been outstripped by the demand for second homes, and this has resulted in the urban population looking further afield. In what for the policy-makers must have been an unexpected consequence of village concentration, deserted houses in non-viable villages, along with their adjacent private plots, began to be bought up by urban residents. This development moved into the forum of public discussion in 1983 when letters were published in *Izvestiya* under the title of 'Houses with Boarded-up Shutters'. Those in favour of allowing the sale of houses to continue argued that temporary occupancy was a good use for otherwise abandoned houses. The counter-argument stressed the illegality of such sales under Soviet law, which only permits the employees of agricultural enterprises and certain other rural residents to hold private plots. As one rural correspondent urged, 'The Land Code must be enforced. If people want to work on the land, let them move to the countryside . . . Don't encourage summer residents who seize state land – prohibit their unauthorized invasion of the countryside!' (*CDSP* 1983d). The official pronouncement on the debate came in the following year in an article confirming the illegality of the transactions and reminding collective and state farms of their obligation under the Soviet Land Code to put all unused land into productive use. People who had bought houses in deserted villages were warned that they ran the risk of falling foul of the law (*CDSP* 1984b: 22). It remains to be seen whether future newspaper columns carry accounts of continuing violations.

Soviet restrictions on research by foreigners mean that the identification of the issues currently at the forefront of rural politics must be based upon the analysis of essentially anecdotal evidence. For the postwar period these sources have hinted at a continuing struggle to get farm management to take the task of rural construction seriously, an expanding but still constrained local authority rôle in rural affairs, and a growing pressure of the urban population on the countryside which has been both exploited and resented by rural dwellers. In the future these issues are likely to remain the same because they derive from the state's attempt to pass financial responsibility for rural development onto farm managements and individuals while maintaining control in the political and ideological spheres. Statistical returns show an upward turn in the provision of housing, education and health-care facilities and increases in per capita wages and the turnover of retail goods in rural areas, especially in the past 20 years. Yet the legacy of past neglect and past planning mistakes has not been overcome. Rural areas are still officially admitted to lag behind towns for a whole range of standard of living indices and, with some villages still without running water, with irregular electricity supplies and cut off by snow for months in the winter, absolute standards remain low. The one advantage that rural dwellers continue to have that is denied urban dwellers is the ability to cultivate their 'private plots' and to raise animals. The presence of these plots with their small outbuildings alongside the collective and state farms' production complexes and administrative buildings gives Soviet villages a distinctive appearance and underlines the fact that elements of the old peasant culture and economy continue to exist alongside the new. The challenge to future Soviet rural policy is whether planning strategies can be developed to take account of them both.

Note

1 Soviet statistics do not record the occupational and social structure of different types of settlement. The 63.5 million figure given for the agricultural population in the statistical yearbook of 1985 refers to farm workers and their dependents working in socialized and unsocialized agriculture. Although the majority of these can be expected to live in rural settlements, Soviet definitions of towns and settlements of urban type allow up to 15% and 25%, respectively, of the population of each to be involved in agriculture. It is likely, therefore, that the population involved in agriculture in rural areas is somewhat less than 63.5 million and the number in non-agriculture-related employment somewhat more than 32.8 million.

References

Balezin, B.P. 1972. *Pravovoi rezhim zemel' sel'skikh naselennykh punktov*. Moscow: Izd. Moskovskogo Universiteta.
CDSP (the Current Digest of the Soviet Press). Published weekly since 1949, Columbus, Ohio, USA.
1983a   1, 15.
1983b   22, 16.
1983c   24, 23.
1983d   42, 21.
1984a   51, 14.
1984b   6, 22.

Erofeev, B. 1970. Pravovoi rezhim zemel' perspectivnykh sel'skikh naselennykh punktov. *Sovetskaya Yustitsiya* **9**, 22.

Gosudarstvennyi Komitet SSSR po Statislike 1987 *Narodnoe Khozaistvo SSSR za 70 Let*. Moscow.

Khodzhaev, D.G. & B.S. Khorev 1972. The conception of a unified system of settlement and the planned regulation of city growth in the USSR. *Soviet Geography* **13** (2), 90–8.

Konchukov, N.P. 1979. *Planirovka sel'skikh naselennykh mest*, 2nd edn. Moscow: Vysshaya Shkola.

Kovalenko, E.I. 1970. *Pravo zemlepol'zovaniya kolkhozov*. Perm: Uchebnik MViSSOb RSFSR.

Kovalev, S.A. 1968a. Problems in the Soviet geography of rural settlement. *Soviet Geography*. *Review and Translation* **9**, 641–51.

Kovalev, S.A. 1968b. Paths of evolution of rural settlements. *Soviet Geography*. *Review and Translation* **9**, 651–64.

Kovalev, S.A. & N.Ya. Koval'skaya 1980. *Geografiya naseleniya*, Moscow: Izd. Moskovskogo Universiteta.

Naselenie SSSR. Spravochnik 1983. Moscow: Izdatel'stvo Politicheskoi Literatury.

Nikishina, Yu. 1971. Rassmotrenie sudami del o narushenii zemel'nogo zakonodatel'stva i samovol'nom stroitel'stve zhilykh zdanii. *Sovetskaya Yustitsiya* **22**, 9–10.

Pallot, J. 1979. Rural settlement planning in the USSR. *Soviet Studies* 31 (2), 214–30.

Pallot, J. & D.J.B. Shaw 1981. *Planning in the Soviet Union*. London: Croom Helm.

Partinya Zhizn' (1965) 23. O rabote mestnykh sovetov deputatov trudyaschizhysya Poltavskoi oblasti 23, 16–19.

Programma Kommunisticheskoi Partii Sovetskogo Soyuza 1961. Moscow: Politzdat.

Pravovie Problemy Rukovodstva i Upravleniya Sel'skogo Khozyaistva v SSSR 1970. Moscow: Nauka.

Pravovoi Rezhim Zemel' v SSSR 1984. Moscow: Akademiya Nauk SSSR, Institut Gosudarstva i Prava, Moscow.

Rogozhin, G.N. 1972. Puti sovershenstvovaniya sel'skogo rasseleniya. *V Pomosh' Proektirovshchiku-Gradostroitelyu. Planirovka i Zastroika Sel'skikh Naselennykh Mest* **9**. Kiev.

SDT (Sovety Deputatov Trudyashchikhsya) (The journal for deputies to central and local councils) Moscow.
 1959 Eshche o planirovanii i zastroike sel'skikh naselennykh punktov **12**, 97–98.
 1964 Dom postroen nezakonno **4**, 7.

Sel'skaya Zhizn' Gazeta Tsentral'nogo Komiteta KPSS (A daily newspaper on rural issues) Moscow.
 1980a Jan 10, 3 'Na ulitse pochtovoi'.
 1980b Jan 11, 2 'Dom na sel'skoi ulitse'.
 1980c Feb 6, 3 Pis'ma v gazetu.
 1985a Jan 13, 2.
 1985b Feb 14, 4 'Rastet, Khorosheet sela'.

1985c Mar 31, 2 Pis'ma v gazetu.
1985d Sept 1, 2.
1986 July 8, 3 Pis'ma v gazetu.
Spektor, M.S. & V.T. Tomilin 1974. *Sotsial'no-ekonomischeskie osnovy pereustroistva sela.* Alma-Ata: Kainar.
Tsentral'noe Statisticheskoe Upravlenie SSSR 1985. *Narodnoe Khozaistvo. SSR v 1984g. Statisticheskii Ezhegodnik. Finansy i Statistika* Moscow.
Tobilevich, B.P. 1979. *Problemy pereustroistva sela.* Moscow: Stroiizdat.

7 The USA

WILLIAM R. LASSEY, MARK B. LAPPING and
JOHN E. CARLSON

Introduction

Any attempt to summarize policies and plans for rural America must be
undertaken with full consciousness of the enormous diversity between
regions of the country, between states, and between communities within
states. National policies and plans certainly have their impact on rural
America, but that impact is invariably differential, depending on the
location.

Careful formulation of rural policies remains exceedingly important in
the USA, despite the widely held perception that we are an 'urban' nation.
Approximately one-fourth (28%) of the US population (65 million people)
live in non-metropolitan areas, and roughly 83% of all governmental units
are primarily rural. These units tend to each be responsible for relatively
small numbers of people, but often encompass geographical areas as large
as some nations (Henry et al. 1986).

CULTURAL AND EMPLOYMENT DIVERSITY

Cultural diversity characterizes rural America. Unlike most other nations,
the rural USA was settled by a wide range of relatively distinct groups of
migrants who displaced much of the native American Indian population.
Native Americans were relocated on isolated reservations which remain
scattered throughout rural America.

The early settlers came primarily from Europe, but migrants later came
in large numbers from Asia, Africa and Latin America. For example, the
English settled much of the Southeast and selected other regions; many of
them became slaveholders and were responsible for the importation of the
black slave population, descendents of whom now inhabit many Southern
rural communities. Germans migrated to many parts of the nation, but particularly to the northeastern and midwestern states (Slocum 1962).

Scandinavians settled in the middle-Atlantic states during the colonial
period, and later became a major segment of the rural population in midwestern and central plains regions. Smaller groups of Europeans from

other countries settled rural communities in many parts of the country (Saloutos 1976).

Communities of the Southwest region were heavily influenced by citizens of Mexican background who have retained much of their culture and language to this day. Asians are a major segment of rural communities in the Pacific Coast region. The Amish and Hutterite religious colonies provide a colourful tradition in many rural regions, particularly in the East and Midwest.

The strong influence of culture continues to play a major part in rural traditions and attitudes. Many communities have a distinctiveness and quality which carries on from their early origins (Salamon 1982, 1985).

THE INFLUENCE OF OCCUPATIONS

Although agriculture, mining, forestry and fishing have traditionally been the primary rural industries, manufacturing is now the largest employer (40%); but government (13%) and trade (17%) are each larger employers than farming (9%) or mining (6%) (see Table 7.1). Forestry, lumbering, and fishing are major employers in a few states but are relatively insignificant nationwide. Roughly 11% of rural people are classified in employment statistics as 'retired'; they are a growing segment of the rural population (Henry *et al.* 1986).

The major occupational categories displayed in Table 7.1 summarize the situation as of 1984, with a distinction between metropolitan and non-metropolitan population distribution. Although 'non-metropolitan' is not fully equivalent to 'rural' as an official US census category, statistics for the two census classifications are generally similar. Using the non-metropolitan

Table 7.1 Population and employment in non-metropolitan counties.

All counties (3067)	Population	Employment
metropolitan (626)	168,302,000	79
non-metropolitan (2441)	64,580,000	21
manufacturing	7,703,000	39 (of rural)
mining	1,115,000	6
farm/ranch	1,115,000	9
retirement	2,115,000	11
government	2,538,000	13
trade	3,228,000	17
other	1,036,000	5
		100

Source: Henry *et al.* (1986: 24–5).

term has the advantage of facilitating comparisons since more data is published on metropolitan and non-metropolitan county distinctions than is available for rural and urban communities (see Fig. 7.1).

THE CONTINUING TRANSITION

Major transitions have been characteristic of the historical process in rural America. Another transformation is currently underway, with highly positive effects for some segments of the rural population and profoundly negative impacts for others. The farm crisis in the mid-1980s illustrates the point.

Advancing technology has increased food and fibre production substantially. This has been of tremendous benefit to consumers (both in the USA and abroad), to suppliers of the new technology, and to those farm units able to finance and utilize the technology. In the process, a high proportion of less prosperous and less informed farmers have been displaced, and innumerable rural communities have suffered decline and a loss of services.

Federal policies intended to help alleviate the problem of highly volatile prices for agricultural products have benefited the well-financed and more technologically advanced farmers and farming corporations, but have not been sufficient to sustain the farmers unable to secure sufficient financing or knowledge to take full advantage of federal policies. The benefits have flowed heavily to those relatively well-off farmers and corporations least in

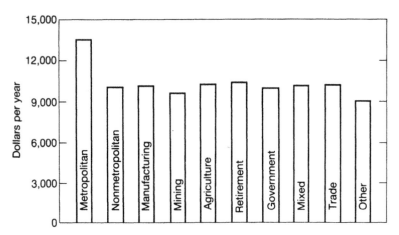

Figure 7.1 Mean per capita income, metropolitan and non-metropolitan counties, 1984.

Source: Henry et al. (1986: 25).

Printed and bound by CPI Group (UK) Ltd, Croydon, CR0 4YY

need of federal programmes, whereas individual farmers and local communities needing outside help have benefited only marginally. International competition with US agriculture has decreased overseas markets and depressed prices, despite policy initiatives to prop up agricultural income (Hansen 1980).

Mining, lumbering and rural manufacturing have been suffering serious decline during the early 1980s. The economic health of rural America has therefore worsened overall, although areas with large retirement populations or government employment have done better – largely because of income transfers from outside the rural region (Henry *et al.* 1986). This has led to certain disparities between metropolitan and non-metropolitan areas, as is illustrated in Figure 7.2. The differentiation between metropolitan and non-metropolitan areas is a long-term trend, however, and has simply been worsened by the decline in rural areas during the early 1980s.

THE RURAL POLICY CONTEXT

Generally speaking, federal-level policy planning for rural areas has three essential elements. Firstly, the US Department of Agriculture (USDA) has been the 'lead' agency in many of the efforts to promote the well-being of that part of the rural sector which in any way could be identified as 'agricultural'. Commodity price supports, credit availability, the provision of research and extension services, and trade policy have been the levers of change which the USDA has sought to utilize to enhance agricultural incomes.

Secondly, in other sectors the US Department of Commerce, through several mechanisms such as the Economic Development Administration, has been responsible for manpower training and infrastructure development programming. The programmes of Commerce and USDA have not always been co-ordinated, and in fact have often run somewhat counter to one another.

Thirdly, and perhaps most importantly, rural policy in the USA has been subsumed under the more inclusive rubric of 'regional policy'. Rural areas have been treated as peripheral or lagging regions, as illustrated by the programme of the Tennessee Valley Authority beginning in the mid-1930s, and the Appalachian Regional Commission of the mid-1960s. There has been a strong bias in much of the policy and programming towards a growth-centre or growth-pole orientation. Urban areas were viewed as the vanguards of development and their prosperity was assumed to 'trickle down' to the rural areas. Demand for resources was expected to increase and jobs were to be created through new manufacturing industries and supporting services (Hansen 1980).

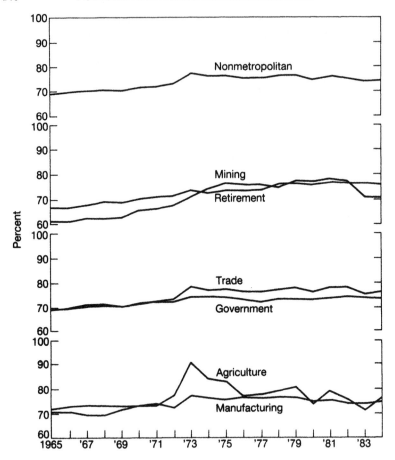

Figure 7.2 Non-metropolitan real per capita income as percentage of metropolitan real per capita income, by type of county, 1965–84.
Source: Henry *et al.* (1986: 27).

A basic expectation was that the provision of better roads, industrial parks, sewage and water systems, and other forms of infrastructure would act as the necessary and appropriate 'magnets' to attract private investments leading to the diversification of the rural economy. This was considered more important than human resource investments such as basic education, job training and other forms of increasing human capacity, largely because of a perception of imprecise results from human development and the expectation of population mobility towards new job opportunities. However, rural people tend to be less mobile than urban people, and there is substantial evidence that one of the most important contributors to mobility is increased education and job skills (Schoening 1986).

Although achievements are apparent in selected communities, the evidence suggests these policies have been less than adequate for most rural regions. A broader development framework seems necessary (Hansen 1980).

This situation is complicated by the continuing industrial redistribution and restructuring process occurring throughout the American economy. Regional variations have been exacerbated, leading to a highly ideosyncratic approach to rural development between states. The fundamental structural changes underway have led to interregional conflict and competition for new or shifting industries, as illustrated by the costly attempts by many states to lure General Motors into selecting them for location of a major new facility (for manufacture of the advanced Saturn automobile, which finally went to Tennessee).

Rural areas of some states, such as California, Florida and North Carolina, have prospered in many respects. But a great many other states in the South, Midwest, Rocky Mountains and West have done poorly.

Major issues in rural regions

JOB DIVERSIFICATION

The illustrations summarized above highlight one of the major issues: the much wider range of employment possibilities now available in some rural communities and continuing limited opportunities in others. Transportation development, modern communication, information technologies and increased amenities in many rural communities (among other factors) have led to a proliferation of new or relocated industries in some regions. Service industries have provided the major employment growth, in response to the demands of increased rural population (Dillman 1985b).

Multiple job holding (two jobs or more) has become the norm for many rural dwellers, both male and female, who are unable adequately to support themselves otherwise. Small-scale farming, fishing, mining, or small business is often supplemented by a job in a nearby urban centre or at a local business. These second (or third) jobs allow individuals and families to maintain a 'rural' lifestyle that would be difficult or impossible with complete dependence on one natural-resource-based occupation.

The dominant sources of primary and secondary employment are in service industries, just as is the case in urban regions. However, the distribution of new employment possibilities is very different from one rural region to another. Selected communities have become prosperous while others have become poverty-stricken (Bradshaw & Blakely 1982).

The occupational distribution of the population in rural areas is illustrated, along with changes over time, in Figure 7.3. The increasing proportion of service employment is particularly noteworthy.

THE PERSISTENCE OF POVERTY

A much higher incidence of low incomes in rural as compared to urban settings has long been characteristic of the USA. The people displaced by advancing technology, and the communities without amenities to attract new employment, suffer serious financial and social deprivation when decline occurs. Pockets of serious poverty therefore continue to exist in regions such as Appalachia and the Rocky Mountain West. Poverty has increased in regions overtaken by decline in agriculture, petroleum production, or other instances of economic downturns (Moland & Page 1982).

The minority population groupings – blacks, hispanics and American Indians particularly – have benefited only selectively from rural economic development and have often been the first to suffer when employment decline occurs. This is in part because of outright discrimination by the white population, but it is also a consequence of inadequate education and socialization to qualify for available jobs. A great many of the poorest rural communities are inhabited primarily by one of these minority populations.

SHORTAGE OF BASIC SERVICES

Despite the overall growth in services as a source of employment, public services in smaller rural communities have failed to keep pace or have diminished. Further, the 'rush to deregulate' American industry has led many private sector service providers, such as trucking, airlines and bus lines, to withdraw from rural areas in favour of the higher-profit-generating urban regions. Health care is an example of the consequences.

Quality and quantity of health care services have advanced immensely nationwide as a consequence of improved technology and an increasing supply of highly trained professionals. Costs have risen very rapidly, to pay for the improved technology, improved professional skills and an increasing array of health services. Rural regions have, however, had great difficulty in securing and maintaining an adequate array of health services (Moscovice & Rosenblatt 1982).

Rural hospitals tend to be small and unable to purchase the new technology as rapidly as urban hospitals. Highly trained health care professionals are hesitant to locate where hospitals are not well equipped for the practice of modern health care. Medicare and Medicaid payments, often a major source of income for hospitals and practitioners, are usually lower in rural regions. Federal policies which attempt to control health care costs have

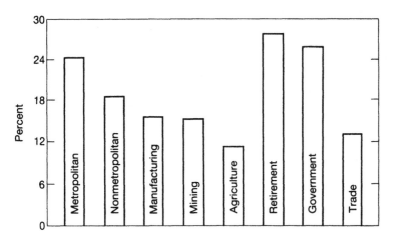

Figure 7.3 Percentage change in service employment, metropolitan and non-metropolitan counties, fourth quarter 1974 to fourth quarter 1984.

Source: Henry *et al.* (1986: 38).

had a more negative effect on rural hospitals and health professionals than in urban settings (Moscovice & Rosenblatt 1982).

Mental health care is often unavailable or of marginal quality, in part because of a value system that does not yet accept mental illness as something to be effectively treated. Mental health professional services and facilities have been difficult to establish and maintain successfully, despite many years of federal and state programmes supporting such activity. Progress has been made, but much less rapidly than in urban areas (Watkins & Watkins 1984).

Health care is but one example of service problems. Many smaller communities are not able to compete with urban centres in supporting grocery stores, clothing stores, automobile sales and service, and a great variety of other commercial services. Recreation possibilities are often limited largely to local social gatherings and television. Schools have been consolidated, thereby diminishing or eliminating one of the primary employers and sources of community cohesion.

Citizens able to travel readily to urban centres have not necessarily been deprived by loss of services. But the older and poorer members of the population often have no choice but to do without (Carlson *et al.* 1981).

CHANGING LAND AND PROPERTY TENURE

The consolidation of agriculture and other resource industries into larger units – often controlled from distant corporate offices – contributes to the decline in local services. As commodity prices reflect a greater degree of international competition, 'land rents', wages and services decline as new sources of production and supply displace American firms from traditional markets for agricultural products. Fewer families are employed in these industries, and as the units grow larger there is a greater tendency to purchase needed supplies outside the local communities where greater economies of volume buying are possible (Goldschmidt 1978, Green 1985).

Changes in land tenure and property ownership also mean a shift of power and influence away from the rural community (Lapping & Clemenson 1983). Leadership in local affairs is often left to poorly prepared individuals. Priorities in the use of available public resources may be disadvantageous to rural citizens in greatest need; services that help sustain a productive environment may be considered unaffordable (Lassey *et al* 1986a).

Land and other property values may decline, as has clearly been the case with agricultural land in many parts of the nation. Given the heavy dependence in rural America on the property tax to finance services, a diminution in quality and variety of public services is observable (Lassey *et al*. 1986b).

THE PROBLEMS OF RURAL GOVERNMENTS

The county, township and village are the principle units of rural government for most parts of the USA. County government is responsible for most open country and is a somewhat unique and enduring institutional arrangement within the various forms of American governance. County commissioners are elected (usually three per county) as both legislators and administrators. They make basic decisions about the allocation of government resources and are then responsible for the direct implementation of many of the decisions – rather than leaving administration entirely to elected or appointed administrators as is the case in towns, cities, some urbanized counties, states, and at the federal level. As a result commissioners are often overburdened and unable to prepare adequately for either good decisions or good administration (Cigler 1984).

Small town and county governments have been impacted dramatically by two major trends: (1) the decentralization process in American government which has shifted increasing responsibilities from higher levels of government to more local levels, and (2) declining or insufficient revenues to finance existing or new services. The decline in federal and state revenue-sharing has had a particularly dramatic impact on many rural jurisdictions. In a 'grants' economy, where jurisdictions compete with one another for limited categorical grant funds, rural jurisdictions are at a distinct disadvantage since they do not have professional grant-writing skills readily available, as is the case in larger and more urban jurisdictions (OECD 1986).

Special districts have proliferated, for services such as fire protection and water supply, each requiring special revenue-generating procedures. This complicates the work of general government, particularly in counties (Cigler 1984, Lassey *et al.* 1986a).

THE BOOM–BUST PHENOMENON

Local government capacity to respond adequately to problems and opportunities is aggravated by the succession of 'boom and bust' cycles characteristic of communities with a narrow employment base and heavy dependence on one or a few industries. Such communities are highly vulnerable to international changes in commodity demands and prices, and often suffer tragically when the primary industry declines or is withdrawn. The centralization in ownership and control of many resources and businesses means that the power of decisions with crucial impacts is often entirely removed from local influence.

Recent examples are plentiful. When international petroleum prices dropped sharply, thousands of rural communities dependent on energy production (primarily petroleum and coal) were very quickly thrown from boomtowns into near ghost towns. Jobs were lost in large numbers. Businesses closed. Services diminished. Real estate became vacant and declined in value. Relatively prosperous citizens were suddenly in search of jobs and/or went bankrupt.

Many rural communities have been selected as locations for new 'high-tech' manufacturing plants. During the recent decline in selected high-tech industries, some of those plants closed, throwing hundreds of people out of work and causing widespread economic and social dislocation.

Although social mechanisms are partially in place to ameliorate such situations (i.e. unemployment compensation) the damage to families and communities remains enormous. Because of the more diverse economies in urban regions, the effects of the bust segment of the cycle are usually not as damaging. Despite the assorted devices (discussed later) for planning and development in rural regions, no mechanism has yet been created for adequately moderating the negative effects of these cycles.

RECREATION, LEISURE AND RETIREMENT AS GROWTH INDUSTRIES

Rural communities in high-amenity regions have been achieving major booms in recreation, other leisure activities and retirement. Urban migration to rural areas for recreation is certainly nothing new, but increased use of highway and air transportation has enabled many places to become much more accessible.

Big Sky, Montana, represents only one of many examples. Famous newscaster Chet Huntley, with the support of the Chrysler Realty Company and other corporate investors, carved a new town from the southwestern Montana wilderness – roughly 45 miles from Bozeman, Montana, to the north and the same distance from Yellowstone National Park to the southeast. The development included a major ski resort, golf, detached homesites, condominiums, restaurants, a dude ranch and numerous other alternatives, all located on former ranching and timber land. It was a highly controversial and (at the time) somewhat unprecedented Planned Unit Development – initiated at a time when rural land-use planning and community development were relatively new concepts in many largely rural states (Lassey 1977).

The new citizens come from throughout the USA, and are generally both highly diverse and affluent. Many of the residents have other homes elsewhere, and a substantial proportion are retired.

Many long-time residents benefited only marginally from the development or were actually displaced. This same situation is characteristic of other new towns based on recreation and leisure as well as other forms of rapid development in many other locations – particularly in the southern, western, and Rocky Mountain regions (Freudenburg 1982).

Retirement communities in rural regions are a relatively new and yet evolving phenomenon, particularly in the so-called sunbelt states. Residence is often limited to individuals beyond age 50 or more. The entire design, physically and socially, is geared to the interests of an older population. Because of the increasing average age and affluence of this population in the USA, development of desirable retirement services is viewed as a potential 'growth industry' in many regions. Yet understanding of appropriate policies and plans for this evolution remains modest.

COMPETITION FOR LAND AND WATER

New town, suburban and urban development is creating a major impact on land and water, and serves as the basis for competition and conflict in many regions. California is the most rapidly growing among American states and illustrates the stress created.

Agricultural and forest land is being consumed at a furious pace by residential and commercial development. This displaces former rural people and industries, while revolutionizing the physical and social environment. Traffic congestion, smog and rapid social change become the norm. Land prices skyrocket, often making agriculture uneconomic relative to competing land uses.

The demand for water to be used by both intensive agriculture and urban development has required the construction of a highly complex system of canals which literally re-route rivers for hundreds of miles. Highly productive Southern California agriculture, and rapidly developing urban areas, are almost entirely dependent upon water imported from the mountain regions of the Sierra Nevada as well as Rocky Mountain states. Needless to say, many rural communities have suffered major dislocations.

Lowering of the groundwater table from irrigation or urban development may cause major environmental, economic and population displacement problems in several parts of the nation. For example, many farmers in the high plains regions of Texas, Colorado, Nebraska, Kansas and the Dakotas have been forced to change from higher-value crops to those with lower values because of groundwater losses. This ultimately has highly destructive impacts on both farmers and local communities. New policies and plans are likely to be needed to deal with this, as well as other water shortage issues (Waterfield 1986).

TRANSPORTATION

Transportation has always been a key factor in rural progress. The closing of railways to rural regions has signalled the beginning of decline, or been a consequence, in many rural regions, particularly if highways are poorly developed. Bus service has been steadily eroding. Airline service has become much more widespread but continues to serve only the larger towns and cities.

The Interstate Highway System has removed much of the traffic from main streets of smaller towns and cities, but has also redirected highway services to those exits which intersect rural areas. Highway construction and maintenance funds have been heavily devoted to this system, and other major highways. This has been a distinct disadvantage to rural regions not located adjacent to these highways. Although the effects of new transportation technology, and decisions about its use, are difficult to quantify specifically, they quite clearly determine a significant part of the future for rural places (Sommers & Zeigler 1983).

CHANGING EDUCATION AND RESEARCH ALTERNATIVES

The Land Grant University system of on-campus education, dispersed research centres and off-campus extension has been a major force in the development of rural America. One or more Land Grant universities are located in every state. Research centres operate in many locations in addition to the main campus and extension offices are active in nearly every rural county. Financial support comes from federal, state, and local sources.

The 'system' has, however, been accused of being, in certain respects, antagonistic to rural well-being. Detractors emphasize the tendency towards support for the status quo when changes appear clearly needed. The focus of research, teaching and extension is judged to be oriented towards corporate and other special interests which use the knowledge in ways antagonistic to the interest of much of the less affluent rural population (Waterfield 1986).

Despite these criticisms, the Land Grant system has the potential for providing a continuing constructive impact in the transition currently under way. Major institutional adaptations may be needed, however, in teaching, research and extension if the potential is to be realized (Warner & Christenson 1984, Dillman 1985b).

The system has become a 'model' for the development of research, education and extension programmes for many other nations, albeit with national and local adaptations. It represents one of the mechanisms for developing rural policies and solving problems that has been uniquely successful in the USA, despite the accusation that it has failed to serve some rural constituencies adequately.

THE IMPACT OF INFORMATION AND COMMUNICATION TECHNOLOGY

New modes of information technology may be in the process of revolutionizing rural America. Ready access to computers and related technologies can (potentially) accomplish many of the knowledge dissemination activities previously performed by classroom teachers, extension agents and other purveyors of knowledge and information.

The proportion of the total workforce employed in the fields of information, knowledge dissemination and various forms of education has reached approximately 50% and may reach 65% by the year 2000. Much of the rural labour force is already employed in these occupations (Dillman 1985a).

Because geography represents no barrier to instant electronic communication, much of the work involved can be performed as well in high-amenity rural areas as in urban centres – which partially explains the dispersion to rural areas of many industries heavily based on information and knowledge transfer. Electronic 'cottagers' are already everywhere in large numbers, doing their work via telecommunication. Education can readily be undertaken at home or in small scattered group settings via video and computer hook-ups.

The information 'revolution' is certainly among the most critical factors in understanding the effects of public policies on rural regions. The geographical isolation that has seemed to hold remote communities back from full participation in national and international life is in the process of being overcome.

Inadequacy of federal and state policies

The federal government has not had a consistent policy towards rural regions in the post-World War II period. Rather, a bewildering array of programmes have been created based on specific problems or population subgroups. The diversity of rural problems and potentialities obviously makes consistency of policy extremely difficult.

Twenty five federal departments or agencies with more than 400 separately funded programmes have had a legislatively required focus on rural development. This great variety of federal agencies with programmes impacting rural areas have been extraordinarily hard to co-ordinate (Powers & Moe 1982).

Roughly every three years the US Congress passes a new farm programme. In nearly all cases, the programme is developed on the basis of the then current 'crisis' in rural America. Although the legislation is referred to as the 'farm program', the contents often deal with a broad range of issues having very little to do with agriculture but with profound implications for rural communities.

The US Department of Agriculture (USDA) is responsible for implementing the 'farm' programme and is largely funded by the 'farm' appropriations bill. Although a high proportion of the appropriated funds are allocated to price supports for a limited range of agricultural commodities, and to food stamps for low-income people, USDA's responsibilities include policies and programmes affecting most rural communities whether or not there is significant agricultural employment. Several other federal departments also have major programmes affecting both agriculture and rural communities (Bradshaw & Blakely 1982).

The Department of the Interior is responsible for the Bureau of Land Management, for example, which controls vast areas of public lands used for grazing cattle, recreation and other uses. Interior's National Park Service has a major impact in more than 300 rural regions where parks are located. The Bureau of Reclamation has major responsibility (with the Corps of Engineers in the US Department of Defense) for managing rivers and streams throughout the nation. Reclamation projects are often the 'lifeblood' for irrigated farming as well as for many other occupations (Carlson et al. 1981).

The USDA contains the US Forest Service, which, along with Interior, manages nearly one-third of the US land area. Many of the nation's most remote communities are dependent upon these public lands for employment, through involvement with lumbering, recreation, water management, ranching and a great variety of other activities generated by the presence of public land.

REGIONAL INITIATIVES

Among the approaches taken by federal policy has been support for major regional planning and development initiatives. The regions have been defined variously, on the basis of river basins, major areas of poverty, areas subject to substantial population growth and environmental impact, and regions under federal jurisdiction such as national forests. Several federal departments have usually been involved, as well as state and local governments responsible for portions of the target region.

The Tennessee Valley Authority (TVA), referred to earlier, is the classic river basin development prototype. The initiative stressed regional planning which involved each level of government and provided multiple benefits for citizens in seven states: Alabama, Georgia, Kentucky, Mississippi, North Carolina, Tennessee and Virginia (Steiner 1983).

The focus of TVA was initially on water as the basic resource for developing a very poor rural region. Flood control, navigation, power production, reforestation, soil conservation, outdoor recreation, manufacture of fertilizer, and other activities, were undertaken to improve the economic and social livelihood of people in the region. Despite its several inadequacies, TVA is considered by many observers to be among the most successful regional policy and planning efforts in the USA (Steiner 1983).

The Appalachian Regional Planning Commission focused on reducing poverty in a thirteen-state region. The governing organization was comprised of representatives from both federal and state governments, with local involvement required through multicounty development districts. In this case the focus was on economic development through the improvement of the physical infrastructure.

Activities have included a wide range of programmes, including a major focus on transportation improvement, vocational education, health improvement, housing, water systems and natural resource enhancement. The assumptions underlying these initiatives were based on the potential of new road systems, for example, to make attraction of industry easier. Experience has demonstrated, however, that transportation routes encouraging entry also serve as routes of exit, for people and jobs (Briggs 1983).

Regional energy planning has a considerable history as a technique for regional planning and has a major impact on selected rural regions. The

Pacific Northwest Regional Power Planning Commission was created in 1934 under the sponsorship of the National Planning Board. It was responsible for assisting state planning boards in preparing state plans as well as regional studies, such as the Columbia Basin Study, which preceded the development of a major natural resource and energy development plan. It served as the predecessor to a succession of energy-planning activities in the region (Olsen 1983).

Regional policy and planning efforts have been supported in most states, often in the form of multicounty districts supported by the Economic Development Administration (US Department of Commerce). Prominent citizens, representatives of local governments, and agency staff, work jointly on projects in which they have a common interest – such as improvement of the health care system, attraction of industry, or recreation development. These efforts have met with varying degrees of success but have undoubtedly been of major benefit to many rural regions (Hansen 1980).

EFFECTS OF STATE POLICIES

Local rural governments operate at the pleasure of the state government. Many of the publically supported programmes in rural communities are created through state government legislation. The responsibilities of county commissioners and other locally elected officials are defined at the state level. Rural planning and community development programmes are in substantial part created and guided by state legislation. State agencies supervise these activities and often have branch offices in rural communities.

State policies for rural areas are often much more coherent and rational, although more limited in scope, than federal policies – in considerable part because state legislators and agencies tend to be much more immediately representative of, and conversant with, rural areas than their federal counterparts. But this varies widely by state.

State governments do not, however, tend to have resources for funding major programmes of the kind supported by the federal government. Thus, much of the funding for rural land-use planning, for example, was appropriated at the federal level, administered by the US Department of Housing and Urban Development, managed by state planning agencies, and passed through to county or town governments for the creation of land-use plans and support of planning staff.

A high proportion of social programmes – for public health, mental health, older citizens, welfare, corrections and others – have been heavily funded (at least until recently) by federal sources. Administration at the local level is usually under the supervision of state agencies. Federal control was predominant at an earlier point, but this is gradually changing as the process of decentralization proceeds.

It is now clear that states and local jurisdictions will have greater responsibility than formerly for the formulation of policies and plans for the social well-being of rural citizens. The Federal government is relinquishing its rôle as the financier and erratic initiator of programmes to resolve rural problems (Powers & Moe 1982).

Most states, meanwhile, have increased their efforts to provide additional guidance and assistance to rural communities. State departments of planning and community development (or comparable units) proliferated in recent decades, usually with substantial assistance from US Housing and Urban Development grants. Social and health service agencies have extended their services deeper into the countryside than heretofore. Field staff have provided direct assistance in such activities as land-use planning, employment services, mental health services and education.

These activities have in the recent past been funded in part by block grants from the federal government or through 'revenue sharing'. Many states have continued the activities following the recent termination of many federal funding sources, although usually on a reduced scale (Cigler 1984).

More localized rural policy development and planning have been evolving as growth of federal programmes has eased or as programmes have been cut. The total resources available to rural areas for such purposes may or may not have grown larger. Enormous variation between states and within states is evident (Dillard 1982).

CHANGES IN VILLAGES AND TOWNS

Some villages or small towns are incorporated and self-governing, whereas others (usually those of small population) are administered by county government. Although many of the services have now shifted to urban locations, the village was clearly the centrepiece of rural social and economic life in earlier historical periods, and remains the centre of social life for much of the population (Johansen & Fuguitt 1983).

The rôle of village and town has been undergoing substantial change. Nearly half of all villages of less than 2500 population declined in population and business activity between 1950 and 1970, although 25% grew more rapidly than the national population growth rate. This pattern changed rather dramatically during the 1970–80 period, when 39% grew in size and only about one-third declined.

The population growth can be attributed to several factors, particularly growth in manufacturing, recreation, retirement, and government services. The evidence is not yet clear for the 1980s, although there is some indication that decline has set in once again because of the general economic malaise in many rural industries (Johansen & Fuguitt, 1983).

Even when population growth has occurred, villages have declined as retail trade centres. Residents travel outside the local community to regional shopping centres for many of their needs.

Whether or not a village government assumes responsibility for significant policy and planning depends in substantial degree on whether it is a centre of a larger regional government unit, such as a county seat or a major institution (perhaps a college or university). In these latter cases the town is very often the location of state and local agency offices for programmes which have substantial influence on rural development.

The politics of rural policy and planning

The lack of national coherence in rural policy is in considerable part a function of the multiplicity of political interest groups with major interest in federal, state and local programmes focused on rural regions. At the federal level the interest groups include (Waterfield 1986):

(a) The House and Senate Agricultural and Interior Committees which prepare and recommend legislation to the full body and oversee the US Departments of Agriculture, Interior and other agencies with important rural responsibilities. They are the focus of activities by assorted powerful lobbying groups.
(b) The departments and agencies of the Executive branch, such as the US Department of Agriculture, Interior, Commerce, Transportation, Environmental Protection, and the Food and Drug Administration.
(c) The 'rural' lobby, which includes both rural and urban interest groups, ranging from the National Farmers Union and Rural America, Inc., which fairly directly represent rural Americans, to largely urban-based groups such as the Sierra Club, representing environmental protection interests, and the National Association of Manufacturers, representing food processors. These organizations employ highly skilled professional lobbyists to work constantly at influencing Congress and Executive agencies.

A high proportion of the lobbyists are lawyers who have had prior experience in some part of the federal government. They tend to be well acquainted as insiders with the way government works, although they may have had little or no rural experience. None the less, they probably have considerably more influence over legislation and programmes affecting rural regions than do rural people (Waterfield 1986).

It is quite clear that rural residents, especially small-scale farmers, have lost influence at state and federal levels, whereas urban residents have gained. The change to a 'one-person-one-vote' system, replacing geographical representation in considerable measure, has meant that the much larger urban population, and its political representation, has gained both in control over programmes and allocation of government resources. Much higher per capita expenditures of federal and state dollars go to urban areas (Waterfield 1986).

Political involvement at the local level becomes more crucial if decisions about priorities are to be affected by rural people and their leadership, as decentralization proceeds. Without deliberate efforts to influence policy, rural regions may secure even less of the public resources than is presently the case (Bradshaw & Blakely 1982).

Elected and appointed local government leaders are better prepared than in the past. They have substantially greater education and specific preparation for their rôles. In earlier historical periods county commissioners were nearly always farmers or ranchers involved in government on a part-time and largely voluntary basis. Small-town councilpersons were part-time business people or professionals. They now tend to be more broadly representative of the full range of rural community population and interests (Poremba & Lassey 1984).

Furthermore, the National Association of Towns and Townships, and the state and national associations of county officials, have become influential forces in the state and federal legislative halls. They have come to realize the need for experienced professional lobbying, and have acted accordingly (Cigler 1984).

Techniques for rural policy and planning

From an historical standpoint the most coherent rural policies in American history may have been devised during the era of President Franklin Roosevelt's New Deal, partially as efforts to overcome the effects of the Great Depression. The National Planning Board was created, and formalized rural planning was initiated on a number of fronts. This included river basin development projects such as the Tennessee Valley Authority and the Columbia Basin Irrigation Project discussed earlier. Several rural new-town projects were initiated, particularly in the eastern states, but also as part of the major river basin reclamation and irrigation projects. The Regional Planning Association of America (RPAA), organized in 1923, was influential in helping to implement rural planning concepts (Lapping 1977, Friedmann & Weaver 1979, Olsen 1983, Steiner 1983).

Many of these projects were multipurpose and multistate, and included a degree of comprehensiveness that has not been attempted since. These early initiatives continue to serve as part of the ideological basis of current policy.

The Land Grant University system continues to have potential for focusing intellectual resources, technical assistance and research data on the policy and plan development for rural regions. The unique linkage with federal, state and local programmes, through the US Department of Agriculture and state agencies with rural responsibilities, continues to serve as a very productive mechanism, although effectiveness varies considerably between states (Browne & Hadwiger 1982).

A period of intense self-examination is now underway within many of these institutions, to determine how the structure can be adapted to fit the emerging American circumstance. Whether or not the revised system will operate more effectively in enhancing rural policy and planning coherence remains to be determined.

By some standards rural policy and planning is coming of age as an academic and practical endeavour. A considerable number of universities now offer graduate degrees in several variations of rural planning, and substantial resources are devoted to research and knowledge accumulation (Deines 1985). Rural policy has been a major focus of several established disciplines, such as agricultural economics, rural sociology, and geography (Browne & Hadwiger 1982, Dillman & Hobbs 1982; Platt & Macinko 1983, Watkins & Watkins 1984). Land Grant universities tend to be the location and support system for much of this work.

Statutory planning issues

The National Environmental Policy Act, which created and sustains the Environmental Protection Agency, may be among the federal statutes with most far-reaching effect on rural regions. Requirements for meeting the stipulations of the Act (and later amendments) have permeated rural policy, planning and development. Protection of rural resources has clearly been enhanced, despite remaining inadequacies of implementation.

The Coastal Zone Management Act has had a similarly positive impact on bringing greater knowledge and forethought to bear on the use of rivers, streams and ocean shores. It has undoubtedly done much to protect important natural resources.

However, the incompleteness of these national efforts is illustrated by our inability to deal with such major transboundary or 'spillover' issues as acid rain, which has devastated plant growth in many regions of the nation. Soil erosion continues to destroy forest, range and farmland; also it pollutes streams and damages fisheries.

Land-use statutes are rather extensively developed in many states, but they have been unevenly effective in controlling or managing the environment and physical structure of rural communities. Some of the statutes have tended to work against rural interests in favour of the 'amenity-consuming' urban population centres (Daniels & Lapping 1984).

The federal government has contributed to the inconsistencies through an assortment of piecemeal categorical programmes for improving local planning programmes, the development of sewer and water systems, and other efforts to improve the physical quality of rural areas. The most recent iteration is the current effort to channel assistance through the states in the form of 'megablock grants' for the improvement of community infrastructure (Deines 1985).

Concluding comments

The succession of problems and shortcomings noted above should not hide the reality of improved conditions in many rural regions. Technological innovation and expansion of employment alternatives have helped to create a new economic and social order. The capacity of local government to manage rural communities has certainly been enhanced. Physical and social services are vastly improved when compared with earlier decades.

These changes have made 'well located' rural communities attractive to industries and employees as places of residence and work. In certain respects, rural communities are at the 'frontier of change and development' as much as urban regions (Bradshaw & Blakely 1982).

The information revolution and other new technologies have reduced the disadvantages as well as the costs of spatial dispersion. Relatively low-cost telecommunications, with connections to the vast potential of computer systems, facilitates the dispersion of manufacturing and services. These developments increase the linkages to the most advanced segments of the nation and the world. The potential for rural employment is greatly increased, as is the potential for improved services.

The possibilities for negative environmental and social impacts are also enlarged, particularly in those locations where there is an absence of policies and plans to deal with the forces of change. Further, remote communities without access to modern telecommunications may continue to remain underdeveloped.

Rural problems such as these will be difficult to overcome, although there is now sufficient experience and capacity at state and local levels (in many states) to mitigate many of the most negative effects. Unfortunately, a shortage of tax revenues to support policy and plan implementation continues to be a serious problem in many states (Lassey *et al.* 1986a, Bradshaw & Blakely 1982).

The concept of regionalizing policy and planning programmes as well as other services has received a good deal of attention. This has proved useful and economical in many locations, particularly in reducing administrative costs and enabling groups of communities to share the expense of employing highly qualified professionals and advanced technology. Counties, or multicounty units created through interlocal agreements, have been quite successful in providing important services – particularly when it is clear that individual and isolated communities are simply unable to provide the necessary financial support (Rogers 1982).

Under conditions of resource scarcity, the potential of co-operative intergovernmental agreements, such as councils of governments, economic development districts, or regional planning organizations, becomes more obvious. The economies of scale involved in providing public transportation, particularly for older people, is a case in point (Kaye 1982).

The rural policy and planning agenda remains full. A current crisis in agriculture and many other rural industries adds urgency to the need for more coherent federal, state and local policies. The economic and social well-being of rural people presently appears to be declining relative to urbanites. Although much of the problem can be blamed on the consequences of international changes in the economic order, the need for new federal and state policies and improved plans demands high priority.

References

Briggs, R. 1983. The impact of the interstate highway system on nonmetropolitan development, 1950–1975. In *Beyond the urban fringe: land use issues of nonmetropolitan America*, Platt, R. and G. Macinko (eds), 83–108. Minneapolis: University of Minnesota Press.

Bradshaw, T. K. & E. J. Blakely 1982. The changing nature of rural America. In *Rural policy problems: changing dimensions*, W. P. Browne & D. F. Hadwiger (eds), 3–18. Lexington, Mass.: Lexington Books.

Browne, W. P. & D. F. Hadwiger (eds) 1982. *Rural policy problems: changing dimensions*. Lexington, Mass.: Lexington Books.

Carlson, J. C., M. L. Lassey & W. R. Lassey 1981. *Rural society and environment in America*. New York: McGraw-Hill.

Cigler, B. A. 1984. Small city and rural governance: the changing environment. *Public Administration Review* November/December, 540–4.

Daniels, T. L. & M. B. Lapping 1984. Has Vermont's land use control program failed? *Journal of the American Planning Association* **50**, 502–8.

Deines, V. P. 1985. Rural area and small town planning. Presentation to the American Planning Association, St Louis, Missouri, September.

Dillard, J. E. 1982. State land-use policies and rural America. In *Rural policy problems: changing dimensions*, W. P. Browne & D. F. Hadwiger (eds), 135–50. Lexington, Mass.: Lexington Books.

Dillman, D. A. & D. G. Hobbs (eds) 1982. *Rural society in the US: issues for the 1980s*. Boulder, Colo.: Westview Press.

Dillman, D. A. 1985a. Social impacts of information technologies on rural North America. *Rural Sociology* **50**, 1–26.

Dillman, D. A. 1985b. Cooperative extension at the beginning of the 21st century. *The Rural Sociologist* **1**, 102–19.

Freudenburg, W. R. 1982. Social impact assessment. In *Rural society in the U.S.: issues for the 1980s*, D. A. Dillman & D. G. Hobbs (eds), 296–304. Boulder, Colo.: Westview Press.

Friedmann, J. & C. Weaver 1979. *Territory and function: the evolution of regional planning*. Berkeley, Calif.: University of California Press.

Goldschmidt, W. 1978. *As you sow*. Montclair, NJ: Allanheld, Osmun.

Green, G. P. 1985. Large-scale farming and the quality of life in rural communities: further specification of the Goldschmidt hypothesis. *Rural Sociology* **50**, 262–74.

Hansen, N. 1980. Policies for nonmetropolitan areas. *Growth and Change* **11** (April), 7–13.

Henry, M., M. Drabenstott & L. Gibson 1986. A changing rural America. *Economic Review* July/August, 23–41.

Johansen, H. & G. V. Fuguitt 1983. Recent population and business trends in American villages. In *Beyond the urban fringe: land use issues of nonmetropolitan America*, R. H. Platt & G. Macinko (eds), 159–174. Minneapolis: University of Minnesota Press.

Kaye, I. 1982. Transportation. In *Rural society in the U.S.: issues for the 1980s*, D. A. Dillman & D. J. Hobbs (eds), 156–63. Boulder, Colo.: Westview Press.

Lapping, M. B. 1977. Radburn: Planning the American Community. *New Jersey History* Fall.

Lapping, M. B. 1982. Upstate: case studies in rural planning – a review essay. *Journal of the American Planning Association* **48**, 387–88.

Lapping, M. B. & H. Clemenson 1983. The tenure factor in rural land management: a New England perspective. *Landscape Planning* **10**, 255–266.

Lassey, W. R. 1977. *Planning in rural environments*. New York: McGraw-Hill.

Lassey, W. R., L. M. Butler & V. Kullberg 1986a. Reactions of county officials to budget cuts. *Rural Development Perspectives* **3**, 29–32.

Lassey, R., L. M. Butler & V. Kullberg 1986b. Public service budget constraints: the perspective of rural officials in the U.S.A. *Journal of Rural Studies* (in press).

Moland, J. J. & A. T. Page 1982. Poverty. In *Rural society in the U.S.: issues for the 1980s*, D. A. Dillman & D. J. Hobbs (eds), 136–42. Boulder, Colo.: Westview Press.

Moscovice, I. S. & Rosenblatt, R. A. 1982. Rural health care delivery amidst federal retrenchment: lessons from the Robert Wood Johnson Foundation's rural practice project. *American Journal of Public Health* **72**, 1380–1385.

OECD (Organization for Economic Cooperation and Development) 1986. *Rural public management*. Paris: OECD.

Olsen, D. 1983. *From darkness to dawn: the evolution of electric power policy in the Pacific Northwest*. PhD Dissertation, Washington State University, Pullman, Washington.

Platt, R. H. & G. Macinko (eds) 1983. *Beyond the urban fringe: Land use issues of nonmetropolitan America*. Minneapolis: University of Minnesota Press.

Poremba, G. A. & W. R. Lassey 1984. *Effects of budget changes on county programs: comparisons among nine Washington counties*. Technical Report No. 1, ARC Project 604, Department of Rural Sociology, Washington State University, Pullman, Washington.

Powers, R. C. & E. O. Moe 1982. The policy context for rural-oriented research. In *Rural society in the U.S.: issues for the 1980s*, D. A. Dillman & D. J. Hobbs (eds), 10–20. Boulder, Colo.: Westview Press.

Rogers, D. L. 1982. Community services. In *Rural society in the U.S.: issues for the 1980s*, D.A. Dillman & D. J. Hobbs (eds), 146–55. Boulder, Colo.: Westview Press.

Salamon, S. 1982. Sibling solidarity as an operating strategy in Illinois agriculture. *Rural Sociology* **47**, 349–68.

Salamon, S. 1985. Ethnic communities and the structure of agriculture. *Rural Sociology* **50**, 325–40.

Saloutos, T. 1976. The immigrant contribution to American agriculture. In *Two centuries of American agriculture*, V. Wiser (ed.), 45–67. Washington, DC: Agricultural Historical Society.

Schoening, N. C. 1986. Federal policies and rural growth patterns. *Forum for Applied Research and Public Policy* Fall, 84–90.

Slocum, W. 1962. *Agricultural sociology*. New York: Harper & Row.

Sommers, L. M. & D. T. Zeigler 1983. Energy change and evolving nonmetropolitan land use. In *Beyond the urban fringe: land use issues of nonmetropolitan America*, R. H. Platt and G. Macinko (eds), 305–12. Minneapolis: University of Minnesota Press.

Steiner, F. 1983. Regional planning in the United States: historic and contemporary examples. *Landscape Planning* **10**, 297–315.

Warner, P. D. & J. A. Christenson 1984. *The cooperative extension service: a national assessment*. Boulder, Colo.: Westview Press.

Waterfield, L. W. 1986. *Conflict and crisis in rural America*. Westport, Conn.: Praeger.

Watkins, J. M. & D. A. Watkins 1984. *Social policy and the rural setting*. New York: Springer.

Zuiches, J. J. 1982. Residential preferences. In *Rural society in the U.S.: issues for the 1980s*, D. A. Dillman & D. H. Hobbs (eds), 247–55. Boulder, Colo.: Westview Press.

8   *Canada*

GERALD HODGE

Rural Canada in perspective

Canada is a huge country, the second largest in the world in land area. Even discounting its Arctic territory, that accounts for 40% of the total, one is left with 6 million km² that is mostly rural. Grasping the texture of this vast space can be done, at best, only partially. We can, however, construct a *mosaic*, even if not fully complete, that will be representative of the Canadian rural scene. The three main parts comprising this portrait are, firstly, the social milieu and its characteristics and tendencies; secondly, the geographical settings of rural activities; and, thirdly, the governmental context within which rural needs are articulated and planning is carried out.

THE RURAL SOCIAL MILIEU

Over 9.1 million people called rural Canada home in 1981.[1] This amounts to over 35% of the total population and, further, represents a gain in both numbers and proportion since 1971. Table 8.1 lists four kinds of rural habitat of which there are, of course, many variations related to their regional location, function and population composition. Concomittantly, the diversity of rural communities carries with it a diversity of development issues.

There are several aspects of these trends that deserve further scrutiny. The farm population today is only half of what it was in 1961; this parallels tendencies in other countries with a modern agricultural sector. However, if we had reproduced the 1976 quinquennial census figures, it would have been seen that there had been a small increase in farm population in the 1976–81 period, the first such increase in three decades. On the other hand, the fastest-growing segment of the rural population is that group that lives in the countryside but is not engaged in farming.[2] People in this group may be retirees or exurbanites, or employed in other rural primary industries such as forestry, fishing and mining, or they may be service providers in the many small towns. Towns, villages and hamlets – of which there are

Table 8.1   Population changes in rural Canada, 1961–81.

	Population (000s)			Average annual change	
	1961	1971	1981	1961–71	1971–81
Total rural[a]					
number	7439	7557	9118	0.1	2.1
% Canada	40.7	35.3	37.5		
Rural farm	2073	1420	1039	−3.1	−2.7
Rural countryside[b]	1661	1946	3098	1.7	5.9
Small towns[c]	3705	4191	4981	1.3	1.9
Total Canada	18 238	21569	24343	1.8	1.3

Notes

[a]   Including the population of all towns and villages between 1000 and 10 000 population in addition to populations defined as rural in the census. `

[b]   Estimated by subtracting the population of all towns and villages with less than 1000 population from the total rural non-farm.

[c]   Including all towns and villages (incorporated and unincorporated) with populations under 10 000.

Source: Census of Canada.

over 9000 – are also attracting new residents at a fast pace. Indeed, since 1971, their growth rates in each of the two five-year periods exceeded that of urban centres (Hodge & Qadeer 1983).

Figure 8.1 illustrates the trends in the share of population of each of the major forms of settlement in Canada. Broadly speaking, there is a renaissance already well under way in rural Canada in all its segments.

What does this resurgence consist of in social terms? One way to grasp what is happening in rural Canada is to review the trends in some familiar social indicators. Table 8.2 presents comparisons for rural communities, small towns and cities for 1971 and 1981. If one reads down the columns of this table it is apparent that rural farm populations are the only ones that differ significantly from the norm. All other rural habitats vary little from their urban counterparts. And, if one reads across the rows, there is a convergence of all settlement types toward a national norm (roughly represented by metropolitan levels). This tendency toward a socio-cultural homogenization has been underway since 1950 according to previous censuses. Extending this kind of comparison for each of Canada's ten provinces reveals the same tendencies. There are clear indications of a *national* Canadian life-style involving rural people as well as the predominant urban population.

Three aspects disturb this homogeneity of Canadian rural communities.

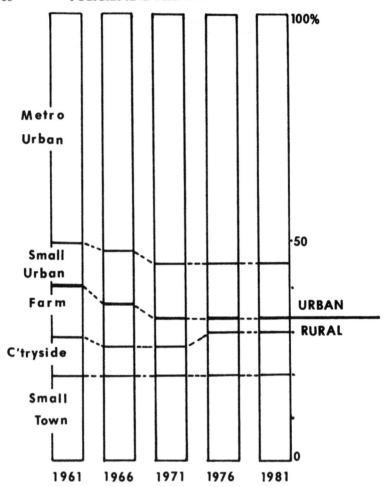

Figure 8.1 Population shares of different settlement types, Canada, 1961–81.

Source: Census of Canada; Hodge & Qadeer (1983).

One is income levels, which tend not only to be lower than for urban areas but also are not as widely distributed (Qadeer and Chinnery, 1986). This, in turn, is related to the much smaller array of occupations, especially fewer professionals and managers working or living in rural communities. Two is the age structure, which tends toward more children and more elderly people in rural areas. Closer scrutiny reveals that farm families and those simply living in the countryside are younger, while town and village populations are older. Three is the presence of native communities, whose living conditions are frequently poor and cultural conditions distinctive.

Table 8.2 Social structural indices of different size settlements, Canada, 1971 and 1981

Settlement type	Household size		Sex ratio (M/100F)		Labour force participation (% population >15 in labour force)		Average annual individual income (% of nation)	
	1971	1981	1971	1981	1971	1981	1971	1981
Rural								
farm	4.3	3.7	116	117	64.2	71.5	70	88
non-farm	3.8	3.2	107	105	50.2	58.2	81	89
Town								
1000–2499	}3.6	}2.9	101	98	}52.9	}59.1	}90	}90
2500–4999			99	98				
5000–9999	3.6	2.9	100	98	55.7	63.0	96	95
City								
10 000–29 999	3.6	2.9	99	97	56.9	63.4	100	97
30 000–99 999	3.5	2.8	98	95	58.0	63.9	99	97
Metropolis								
100 000–49 999	3.3	2.8	}97	94	61.0	65.3	106	99
over 500 000	3.2	2.7		95	59.1	68.2	112	109

Source: Census of Canada.

These aboriginal peoples' future lands are a subject of national debate.

In prosaic terms, if one strolls down the main street of most small towns in Canada one will see the evidence of national systems of marketing in the supermarkets, fast-food outlets, automotive establishments, and so forth. The welcoming signs of national service clubs on the edge of town announce ties to national cultural interests, as do local chapters of the Cancer Society and the Women's Institute. A glance at the rooftop or front-lawn TV dish antennas shows how securely rural people are 'plugged in' to national and even global society.

Not only are rural areas enmeshed in national systems for the production and distribution of goods, services and culture, they are also linked to all variety of public programmes and institutions. For example, laws are relatively uniform and public programmes usually common in all parts of the country. There has been considerable improvement in the provision of social services and physical facilities throughout the countryside and small towns of rural Canada over the past twenty-five years. Probably most important have been the programmes to provide all-weather roads and electric power on an almost ubiquitous basis to rural residents. Numerous federal programmes are also part of this public 'life-support' system which affects the fortunes of rural communities.

What has just been described may be termed the 'facilitating mechanisms', the means that permit choices to settle (or remain) in rural areas to be carried into effect (Sternlieb & Hughes 1977). Thus, the rural population growth and small-town persistence are supported by the mediums of communication, transportation, social services and new commercial structures. But there must be not only the *means* to produce widespread settlement changes in rural regions, there must also be the *inclination* of people to employ the means – to 'want to' move, make different housing choices, etc. These inclinations are confirmed in the choices of people to seek or maintain a rural or small-town habitat as portrayed in the census figures.

The point to be emphasized here is that 'urbanism as a way of life has spread beyond urban areas' (Qadeer 1979). Rural Canada is now firmly part of the modern industrial–commercial–cultural milieu and has problems, expectations and needs not dissimilar to those of urban areas. Thus, planning and development goals are substantially the same as between urban and rural parts of Canada. Differences arise, however, in the implementation of those goals because rural communities differ in nature, size and stage of development from urban ones.

Equally important is that Canadian rural communities differ amongst themselves as much as they differ from urban places. Wirth (1938), long ago, identified three dimensions by which to distinguish communities – size, density and heterogeneity. Rural communities are small, low in density and homogeneous compared to urban ones. In general, this affects their planning programmes and solutions. In actual communities the combination of these factors obliges often quite specific planning approaches even amongst neighbouring communities, much less those a continent apart. A small settlement of ex-urbanites on one of British Columbia's Gulf islands is a world apart from a Newfoundland fishing outport or a Mennonite community in Saskatchewan's grain-growing area, even though each may be the same size. The provision of transportation, health services, or housing is likely to call for custom-made solutions to their planning problems.

THE GEOGRAPHICAL DIVERSITY

The diverse social milieu is matched by the geographical diversity of Canadian rural communities and, in many ways, they are interdependent. The prevailing physiography affects their communications, economic base and life-styles, to name only three obvious links. Geographic differences ensure the uniqueness of rural places. Yet within this diversity are a number of geographic domains whose attributes confer common features on their constituent rural communities. Rural Canada is comprised of five major types of geographical areas:

(1)   Agricultural regions,
(2)   Urban fields,
(3)   Mature resource regions,
(4)   Frontier regions,
(5)   The North.

On a map of Canada (Fig. 8.2) the first four categories recur and overlap each other in the southern half of the country; they define essentially *functional* differences among rural communities. The North is a special rural region by virtue of its population, environment and stage of development. Each is discussed briefly below in terms of its distinctive development and planning problems (see also Table 8.3).

Agricultural regions   Agriculture is conducted in each of the ten provinces. It is the dominant land use in four: Alberta, Saskatchewan, Manitoba (the Prairie provinces) and Prince Edward Island. Production tends to be specialized in the various agricultural regions according to climate and soil capabilities, with the productivity high and mechanization in widespread use. This is true whether the crop be grain in the Prairie provinces, soft fruits in British Columbia, grapes in Ontario or potatoes in New Brunswick. The family farm is the prevailing production unit, with residence on the farmstead. Towns and villages are widely distributed and all such regions have good roads, electric power and telephones on a ubiquitous basis. Schools and hospitals are provided extensively, although on a consolidated basis often requiring extensive travel. The personal motor vehicle is the chief means of transportation. The problems of these regions are, if anything, attributable to their success. Common needs are the maintenance of community life and the continuation of basic health and social services in the face of a highly rationalized agriculture requiring fewer farmers. A paradox is that the elderly populations are showing less tendency to leave, thereby inducing new needs in shelter and services in the small towns they favour for retirement. In Manitoba and Saskatchewan towns and villages in 1981, the census showed the elderly comprising close to 25% of the population (compared to the Canadian average of just under 10%).

Urban fields   About 40% of Canada's farm population live within the 50 km surrounding larger and medium-size cities (Russworm & Bryant 1984). Farmland in these situations is subject to the forces of urbanization, leading to land abandonment, land conversion and part-time farming. This, of course, is indicative of a whole set of problems in the rural parts of urban fields that derive from the competition for space by many different uses and users: full-time farming, hobby farms, rural subdivisions, second homes of city dwellers, active recreation, regional parks, domestic water

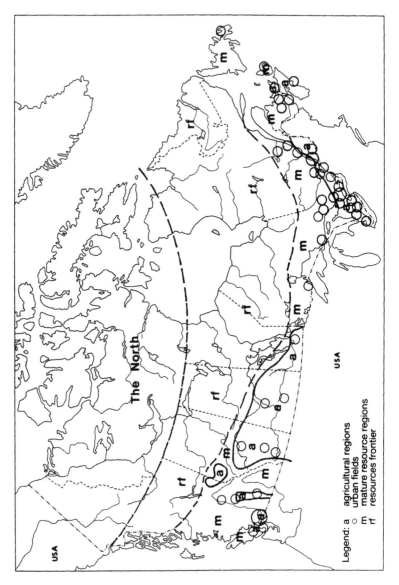

Figure 8.2 Rural regions of Canada.

sources, amusement parks, sand and gravel quarries, shopping centres, and so forth. The surroundings of cities house the greatest number of rural dwellers, have the greatest population densities, and are growing fastest. Thus, their planning is dominated with issues of transition and adjustment between urbanization and agriculture, newcomers and oldtimers, development and the environment.

Mature resource regions   Many regions received extensive development in the exploitation of their mineral, timber or fishing resources; some date from the earliest settlement of Canada. Viable agriculture was limited to small pockets due to the rugged terrain and/or intemperate climate. Towns and villages grew and land was taken up for residence, individual woodlots and marginal farming, yet urbanization never came to these regions. In some, the resources were depleted (New Brunswick's forests) or became obsolete (Cape Breton's coal); in all, agriculture diminished in importance in the national economy.

Settlement persists, however. It can be characterized as small in scale, low density and physically disjointed. Poverty is not uncommon where resources have been depleted. And fluctuations in demand for resources affect the regularity of income in all such regions. Local resources are meagre for providing services and, in their isolation, communities are not easily incorporated into larger service systems. More diversified economic development and upgraded public services and facilities top the planning agenda. Native Indian communities are commonly found in this type of rural region.

Frontier regions   The northern parts of most provinces are the locale for newer resource exploitation – sources of minerals, oil, pulp and paper, and hydro-electric power. Single-purpose communities with limited land links to older settled regions to the south are the norm. Intraregional transport is a rarity. These newer projects are often very large in scale; for example, the multibillion dollar Tar Sands plants of northern Alberta and the James Bay electric power development in northern Quebec. Environmental damage and dislocation of native communities are not infrequent by-products of these projects. Isolated new towns have problems of population turnover, family stress, and limited facilities and services.

The North   Canada's two northern territories comprise over two-fifths of the entire country and yet contain, perhaps, 125 communities of which only two barely exceed 10 000 population. Three-quarters of the people are of Inuit (Eskimo) or Indian descent; the majority of them still live off the land (e.g. hunting animals for furs, meat, clothing) and are supplemented by public assistance. Rich ore deposits and vast reserves of oil and natural

Table 8.3 Rural regions of Canada and their characteristics.

Types of rural region	Area km²	%	1981 Population No.	%	Representative areas	Typical planning problems
agricultural regions (excluding urban fields)	480 000	5.2	1 600 000	6.6	southern portion, Prairie Provinces, Okanagan Valley (BC), SW Ontario, Eastern Townships (PQ)	provision of services; transportation; housing the elderly
urban fields (excluding urban areas)	320 000	3.5	3 900 000	16.0	area within 50 km of 52 cities over 40 000 population	agricultural adjustment; land-use conflict over recreation uses; changing rural communities
mature resource regions	1 530 000	16.7	2 700 000	11.0	Newfoundland, New Brunswick, NE Ontario, Northern Vancouver Island	poverty; depleted resources; provision of jobs and services
frontier regions	2 900 000	31.6	500 000	2.1	Labrador and northern portions of most provinces	isolation; housing; population turnover; transportation; poverty of native population
the North	3 780 000	40.9	70 000	0.3	Yukon and NWT	isolation; native land claims; transportation; environmental impacts

					central city and suburbs of 52 largest cities	land-use conflicts; provision of housing; transportation
major urban	190 000	2.1	15 570 000	63.9		
Canada	9 200 000	100.0	24 340 000	100.0	n.a.	n.a.

Sources: *Canada year book*; Census of Canada; Russworm & Bryant (1984).

gas are the contemporary lure of the North for corporations and governments, replacing the individualistic gold rush and fur trading of the past.
New development is constrained by aboriginal land claims which were
never settled previously. The North is one of the world's last frontiers that
has, on the one hand, an intemperate climate and, on the other, a very deli-
cate ecosystem. The extremely high capital intensity of projects in this
region, along with unresolved native issues and transportation, cloud the
future of the North, as they always have. In the meantime, communities
will continue to be small, poorly serviced, isolated and mostly poor.

THE GOVERNMENTAL CONTEXT

Canada is a federation of ten provinces that, collectively, have assigned
some powers to the national government. The provinces control the use of
all natural resources, including land, within their respective boundaries.
The two northern territories are within the national government's domain.
The latter government also has residual powers in areas of transportation,
water resources and economic development. Further, all the provinces
have, to a greater or lesser degree, devolved some powers onto local
governments. The most important of these for rural development are those
dealing with the provision of services to property (e.g. roads, protective
services, recreation) and the planning and administration of land use.

Rural regions, except the North, function within this division of powers.
The province has the key rôle: it can *delegate* its municipalities to respond to
planning problems and it can *invite* participation from the federal government in solving problems within its boundaries. Thus, as long as problems
occur wholly within municipal jurisdictions or within the province there
are likely to be statutory means to respond to them. If they overlap either
the municipal or the provincial realms, special means are required such as
regional planning agencies or federal–provincial agreements. Needless to
say, few rural planning and development problems fully respect political
boundaries or functions.

The rural jurisdictional situation may, perhaps, be better grasped by
referring briefly to the problems of the several rural regions described
above. Agricultural regions experience problems at all levels from those
occurring in the community to those occurring with regard to the transportation of crops for international export. The latter is clearly a federal
government responsibility. The provision of local health and social services
involves provincial government ministries utilizing province-wide standards for their distribution and quality at the community level. Needs for
senior citizen housing are met through initiatives beginning at the local
level but funded by both senior levels of government, with the province
deciding on the actual allocation to one community or another. Local

governments in agricultural regions, and in other rural regions as well, generally lack the financial strength to solve many of their problems; their tax base is almost entirely derived from taxes on real estate. Most receive subsidies from the province.

In urban fields the problems overspill municipal boundaries and their solution is constrained by the lack of adequate, or in many instances any, institutional structures to reconcile such issues. Metropolitan governments exist for many of the large cities, but their jurisdiction is usually limited to an area much smaller than that over which urbanizing effects are felt. Regional municipalities are used in some provinces and others employ regional planning commissions to deal with rural–urban problems. There is, in short, no uniform system of problem-solving and planning for urban fields; most such arrangements are *ad hoc* and limited to a single function such as transportation, land use, water supply, etc. Provincial sanction is required in establishing any intermunicipal bodies.

In mature resource regions and frontier regions local government systems are often completely lacking, except for major towns. Dispersed rural settlements, which are often the norm, are served, if at all, by provincial departments. Ministries offer programmes in health, welfare, agriculture, housing, etc. as, for example, subsidies for doctors to practise in rural areas. The planning for these programmes seeks to promote provincial objectives, and, as Qadeer (1979) further notes, local priorities receive little consideration and impacts of these programmes are 'often as disruptive as beneficial'. This top-down approach, coupled with meagre local resources, generates a high degree of dependency on the part of rural communities. All rural regions suffer this same dilemma.

Rural planning from the top down

Rural areas in Canada are very *dependent* upon senior governments, as the preceding discussion of the governmental context indicates. This is true whether the need is to provide the stimulus for economic development, basic social services or the funding for physical infrastructure and community services. This stems from the persistent fact that rural areas do not have sufficient of their own resources – financial, administrative, technical or human – to plan and undertake their own development. Most are removed from areas of diversified economic development, and their staple resources (agriculture, forestry, mining, fishing) are, at best, economically slow growing. Only in rural parts of urban fields are development opportunities more extensive, but even these rural communities are constrained in their planning and development capabilities.

This dependency position reflects not only the disparity in resources but also the outlook of senior governments. Rural areas are seen as the source of raw materials for the dominant urban industrial society in Canada or as residual situations in the process of the nation becoming urbanized and industrialized. Rural planning by senior governments is undertaken mostly to promote national and provincial objectives (Qadeer 1979) which, in general, are urban in their orientation and priorities. The rural planning that has emerged in this kind of context tends to be disjointed, inconsistent in its application, paternalistic and insensitive to local impacts.

The top-down approach to rural planning takes a number of different forms depending upon the objectives and outlook of the senior governments. It may stem from policies aimed at the welfare of all citizens and communities, as in providing basic health services and highways. Or it may stem from policies aimed at improving the economic well-being of residents of older natural resource regions. In all, there are six different, but not infrequently overlapping, forms of rural planning initiated by Canadian senior governments:

(1)   planning for rural redevelopment;
(2)   urban region planning;
(3)   conservation of natural resources;
(4)   community land-use planning;
(5)   distribution of basic services;
(6)   mega-projects planning.

Each will be described briefly with the emphasis placed on the intended and unintended effects on rural people.

PLANNING FOR RURAL REDEVELOPMENT

The most extensive and significant rural planning activity began in 1961 with the passage of the federal Agricultural Rehabilitation and Development Act (ARDA). Through this legislation the federal government undertook to work with provincial governments at relieving rural poverty in agricultural (and forestry, fishing and mining) regions. As cities burgeoned and their secondary industries accelerated the economy, non-urban regions with persistent poverty began to demand a share of the prosperity. These were, notably, the mature resource regions of the four Atlantic Provinces and Eastern Quebec. Studies showed these areas to have many of the classic symptoms of economic underdevelopment: poverty, illiteracy, poor housing and public infrastructure, obsolete resources and inefficient technology. Under ARDA low-income rural regions were designated to receive grants (e.g. to promote farm enlargement), services (e.g. assistance in establishing community pastures) and facilities (e.g. improving farm market roads).

About 40 rural regional planning efforts were undertaken in the 1960s under ARDA from southeastern British Columbia to Newfoundland. Although it started with a people-oriented criterion of need, its means were largely investments in economic activities and physical infrastructure rather than in social services, education or housing. The planning did not usually proceed according to a published plan or with permanent planning staffs. There were federal provincial task forces, development corporations, and various funding agreements depending upon the province involved. The degree of participation of local communities and citizens in the planning also varied considerably.

One particular effort under ARDA deserves special mention – the Newfoundland Outport Resettlement Program. The several versions of this programme aimed at relocating many of the numerous fishing villages situated along the rocky and variegated coast of the province so as to provide the population with basic services and job opportunities in more accessible locations. Of the approximately 1400 outports, as they are called, about half are located on isolated headlands, bays and islands. They are mostly very small (50–200 people) and only accessible by water. Resettlement, facilitated by grants to households to cover the cost of their move to specified larger centres, was voluntary (Copes & Steed 1975).

It is variously estimated that the populations of about 300 outports were resettled. However, just as many refused to take up the option. Resettlement caused splits in families and communities. In addition, there was often no prior consultation with communities planned for relocation, and the receiving communities often were not as well equipped as promised (Matthews 1976). The presumptuousness of the programme's planners has been widely criticized and, down to this day, it is considered a blot on the otherwise thoughtful ARDA effort.

With the end of the 1960s, ARDA was absorbed by a new ministry primarily devoted to economic development of rural and resource regions through the creation or expansion of secondary industries. This ministry, the Department of Regional Economic Expansion (DREE), provided grants and loans to firms to locate in designated regions and also grants to local and provincial governments to improve infrastructure to help attract firms. A growth-centre strategy was employed with rural areas expected to benefit through 'spread effects'. Since the early 1980s, these broad-scale efforts at rural redevelopment have been discontinued more because they lack a political constituency than because the need has gone.

URBAN REGION PLANNING

Immediately following World War II, Canadian cities, like those elsewhere, began to experience extensive growth. Housing, shopping facilities and factories expanded into surrounding rural areas. New demands were placed on agricultural land, water resources, and rural open space and new planning instruments were established to try to deal with the 'exploding metropolis'. The earliest were the regional planning boards in British Columbia in 1949 that were established by the province for the areas surrounding the two major cities, Vancouver and Victoria. These regions encompassed broad rural hinterlands up to four times larger than even today's metropolitan development covers. Rural municipalities were assisted in planning their land use, conserving agricultural land, and limiting 'urban sprawl'.

The Province of Alberta followed suit in 1950 with district planning commissions for Edmonton and Calgary, and later for all major centres. Nearly 60% of the area of the province is covered by these commissions. Although formed to deal with urban containment, most of the commissions perform land-use planning functions for constituent rural municipalities. The Calgary Regional Planning Commission, for example, has developed a format for small-town plans in which all parts of the plan required under provincial statute are printed on one large, poster-size sheet. This is much more 'appropriate technology' than the cumbersome, legalistic plans that have typified rural community plans in most other provinces. The Oldman River Commission in southern Alberta has undertaken much more than land-use planning in its district. Planning for libraries, solid waste disposal and ambulance services in rural parts of the district are among their activities.

Most planning in the rural parts of urban regions tends, however, to be oriented to the needs of city dwellers and city businesses. Rural areas are planned in order to guarantee water supplies, accommodate regional transportation facilities, provide parks and open space for city dwellers and, more recently, be the repository of hazardous wastes. Planning for the protection of food-producing lands from urban encroachment is the one effort that also coincides with the aims of many rural dwellers in urban regions.

CONSERVATION OF NATURAL RESOURCES

The use of the natural resource base of rural regions has increasingly been constrained in the past two decades by environmental and land conservation policies. Concern has become pervasive over the decreasing supply of agricultural land, the dangers of over-harvesting fish and forests, and the pollution of water supplies. In the public interest a variety of initiatives have been taken by senior governments that affect the life and livelihood of Canada's rural dwellers.

The supply of agricultural land in Canada is a source of national concern. Even in such a large country, only 8% of the land is suitable for crops; further, much of the best land is located near to cities. For example, urban development in the Niagara Peninsula threatens the valuable soft fruit industry (it is one of the few places these crops can grow in Canada). Regional planners are defining firm boundaries on the extent of new urban development in this part of Ontario with some success. The provincial government of Ontario has also published *Foodland guidelines* setting out criteria for municipalities to employ in deciding upon new land-use proposals in areas with good soils. The British Columbia government established the Agricultural Land Commission in 1973 with extensive powers to designate Agricultural Land Reserves on which no urban development can take place. This latter agricultural 'zoning' has proved quite durable and, generally, has the support of the farming community because it sustains rural communities and life-styles.

Beyond conserving farmland there is the need to ensure how it will be farmed. Farming practice, in an era of high mechanization and sophisticated techniques, has the potential of creating unwanted environmental effects. The use of pesticides and fertilizers, the disposal of animal wastes, and the noise of machinery may have deleterious effects on neighbouring properties of farmer and non-farmer alike. Water bodies serving larger populations may also be affected. In Ontario the *Agricultural code of practice* is a set of detailed specifications regarding such hazards. British Columbia, another province where the juxtaposition of farm and non-farm land use is common, is considering adopting similar regulations. Also, three provinces have enacted 'right to farm' legislation – New Brunswick, Manitoba and Quebec – two other provinces are contemplating it. These statutes prevent farmers from being sued over farming practices that may offend neighbouring property owners.

The Province of Saskatchewan (Canada's major wheat producer), alarmed at the rapid ageing of its farm families and the high cost of entry into farming for young farmers, instituted a programme to assist the transition to a younger farm population and maintain the favoured 'family farm'. Under this programme the province could purchase farms of farmers wishing to retire and re-sell them to young farmers on favourable terms.

Natural resource harvesting practices in forestry, fishing and mining have also come under increasing scrutiny and control. Limits or quotas are used to prevent over-harvesting (in farming, quotas are used as well, but to prevent over-supply). In some rural regions, such as Quebec, the Atlantic Provinces and British Columbia, mineral, fish and timber resources occur in close proximity. Harvesting practices in one not infrequently causes problems for another, e.g. sedimentation and chemical leaching affecting water bodies and fish stocks. When controls are instituted, the livelihood of rural dwellers is affected and rural communities disturbed. Seldom does this enter into natural resources planning in Canada at the outset.

COMMUNITY LAND-USE PLANNING

All the provinces have legislation that permits rural communities that are incorporated as municipalities to plan and regulate their physical environment. These statutes, initially modelled on the 1909 British Town and Country Planning Act, prescribe the format for local land-use planning: the type of plans to be prepared, the permitted land-use regulations and the administrative arrangements to be used. Provincial Planning Acts had their beginnings in the concerns over the ills of rapid, large-scale growth of *cities*. The planning needs of rural communities are, however, different in scale, intensity and pace of development. The prescribed formats of Planning Acts are generally incompatible with the land-use planning situations of most rural municipalities (Hodge & Qadeer 1983).

In most rural communities in Canada, land development tends to be small in scale (perhaps 15–20 new dwellings per year), not very intense or diverse (mostly housing with occasional shops or public buildings) and varied in pace (with some development one year but none the next). Thus, neither elaborate plans nor conventional land-use regulations are appropriate for rural communities. On the other hand, towns, villages and countryside communities see their development in terms of a set of specific problems arising in their particular setting. An insightful report prepared for the Newfoundland government recommended that rural community plans should use the problem-solving approach (Newfoundland Department of Municipal Affairs 1968).

Provincial Planning Acts also assume that all communities are capable of establishing a workable planning function. However, rural communities are small and the job of making planning policy and implementing it usually falls to a few hard-pressed officials. Small municipalities frequently have administrative staffs of less than five people (and some may only work part-time), such as the building inspector, roads superintendent and town clerk. Rarely does one find a planner on staff in communities less than 15 000 in population. This has often led to a disdain of planning edicts from the province: Planning Acts may be disregarded and plans, where they exist, may be ignored. Lastly, large numbers of rural communities are not eligible in any case to undertake land-use planning because they are not incorporated.

Provincially sanctioned land-use planning creates an illusion of local communities being able to influence their development. Yet local plans and by-laws must receive provincial approval. Further, there is almost no co-ordination of this phase of provincial activities with those that it undertakes for farmland preservation, environmental protection, economic development, and so forth, for the same communities. At the same time, the problems many rural communities need to solve – the lack of water and sewer systems, the need for senior citizen housing, refurbishing 'Main Street', improving sidewalks and street-lighting – are not addressed by this type of planning.

DISTRIBUTION OF BASIC SERVICES

All provinces have programmes to provide basic services and facilities to urban and rural communities alike: health and social services, community facilities, education, water and sewerage, electricity, telephones and highways. As indicated at the outset of this chapter, the ubiquity of many of these services in rural Canada is responsible for a large part of the recent upsurge in rural population growth. The last three named 'hard' services are particularly important in this regard. The picture is more mixed, and frequently not as sanguine, in regard to 'soft' services in rural areas. These programmes are often flawed by the failure to acknowledge differences among rural communities or by shortfalls in the delivery by senior government agencies.

Concern is expressed many times by rural communities over the *standardized* approach of provincial and federal programmes. As Baker (1960) noted nearly three decades ago, 'what one community can do very easily, another may find very difficult'. There are differences in the available resources, leadership and attitude of communities that may influence their ability or inclination to access programmes. A programme for improving recreational facilities in small communities in Ontario, for example, required the preparation of a 'community recreation plan' in order to apply for grants. Programmes for providing public utilities, libraries and housing usually call for costs to be shared and many rural communities cannot afford their portion. Communities with high proportions of the elderly may, for example, feel inhibited to devote local tax money to programmes designed to expand the size of the community.

After entering into programmes of senior governments, rural communities may then find themselves frustrated many times by delays, lack of co-ordination, etc. The sheer number of agencies employed to deliver policy is daunting. Field studies have shown that Nova Scotia and Manitoba rural communities, in order to serve nine common activities and functions, had to deal with 16 different provincial agencies in the former province and 13 in the latter (Hodge & Qadeer 1983). Duplication, interdepartmental conflict and competition among programmes have been readily visible to rural residents (cf. Pross 1975).

The major flaw in policies and programmes for basic services being delivered to rural areas is that they are usually not designed with *rural* communities in view. A significant exception to this was the 'Stay Option Policy' of the Government of Manitoba in the early 1970s. This policy aimed at providing rural communities with a level of well-being in housing and social services such that rural people would not feel forced to migrate to the city: they could stay in their communities, if they chose. It featured the co-ordinated delivery of provincial programmes to individual communities as well as the general upgrading of service levels in rural areas.

MEGA-PROJECT PLANNING

Canada's development, historically, has depended upon major capital-intensive projects, from the transcontinental railway to the St Lawrence Seaway. Today, the emphasis is less on transportation and more on energy resource projects. Although the aim of the latter projects is to satisfy the needs of metropolitan areas, most of the direct impact is felt in rural regions. Construction workers throng into these areas for a few years, straining housing and social services. The aftermath may feature a segment of newcomers, perhaps in their own new town, and local pollution problems. Dealing with these impacts is often left to the local community and is seldom addressed in the planning for mega-projects. Two examples will illustrate the dilemmas faced by rural areas.

Southeastern British Columbia has enormous soft coal reserves that have been mined since the turn of the century. Around 1970, a multinational corporation determined that new markets of unprecedented size existed in Japan if the coal could be obtained by strip mining rather than from the old underground mines. Permission was granted for the open-pit mines which would ship more coal in the first 7 years of the project than had been shipped in the previous 70 years. Two old 'company towns' were to be shut down and a new town built some distance away. This led to numerous relocation problems for residents and businesses: higher costs for housing, insufficient relocation assistance, traumas over leaving familiar surroundings, etc. In addition, pollution of streams and rivers became a major problem as the new mining techniques removed base areas of overburden.

The Strait of Canso in Nova Scotia was selected for a massive petrochemical complex by the federal and provincial governments, also around 1970. Port Hawkesbury is the main centre in the area and the only one with a local government. It was urged to prepare for a nearly tenfold growth in population to about 10 000 by the province's planners (Pross 1975). Not only was this forecast far short of the town's actual growth (3900 in 1981) but an additional 3000 persons were allowed to sprawl into adjacent unincorporated areas where no provision was made to provide necessary services for them. Port Hawkesbury, meanwhile, undertook huge borrowing for sewers, roads, water system and a community centre. This rural project area also had increased social costs due to social instability, human dislocation and the strain on social services and facilities.

Public outcry over the social and environmental impacts of the proposed multibillion dollar Mackenzie Valley oil and gas pipelines led to a notable effort at pre-planning a project (Berger, 1977). The Commissioner, Thomas Berger, took great pains to listen to and understand the concerns of inhabitants of several dozen native communities in the region. These involved the social intrusion of the construction, the environmental disruption of traditional sources of food such as caribou and fish stocks, and participation in the region's economy in the longterm. Berger's report recommended a ten-year delay in beginning the project to allow native peoples time to prepare for the impacts. The federal government heeded the advice.

Community-based rural planning

Rural planning from the top-down in Canada is so frequently flawed because it is not appropriate to rural situations. Basic dimensions of rurality must be respected such as the small size of communities, the low density of activities and facilities, the limited capacity of human and financial resources, the strong social networks, and the slower and less regular pace of change. These dimensions result, in turn, in distinctive needs from one rural community to another. The most likely place to obtain an appreciation of these needs is from the rural community itself. Since most top-down planning does not start here, its expectations for success are often dashed.

By contrast, community-based rural planning in Canada shows a high rate of success. Although not actively encouraged by senior governments, local initiatives to solve rural community problems are widespread. There is a great diversity in these efforts and their organization, as befits the diverse rural milieu, and their focus is highly localized. The number of such efforts can only be guessed at, for they do not usually derive from formal programmes and they tend to eschew collateral associations. It is not unreasonable to assume there are a few thousand in the form of co-operatives, local development corporations, voluntary service associations, regional planning agencies and social planning councils.

The variegated texture of local efforts at rural problem solving may be demonstrated by several examples from different parts of the country. Four broad categories will be used to identify substantive areas rural people are dealing with on their own: (1) local economic development, (2) delivering social services, (3) mobilizing rural resources, and (4) rural regional planning. It should be emphasized that within these categories there are scarcely two alike, but all are characterized by the application of planned approaches and community involvement.

LOCAL ECONOMIC DEVELOPMENT

There is a new surge of community self-reliance efforts in the small communities of Canada. It is the same spirit that led early settlers to build roads in lieu of taxes and of those who, later, established rural telephone companies and producers' co-operatives in grain growing, logging and fishing communities. It arises in response to shortfalls in traditional strategies of economic development: waiting for investments from outside that frequently do not materialize. Or when they do, bring unacceptable development, pollution and stress. Thus, one finds dozens of community economic development projects in rural regions from coast to coast being initiated in order to gain a greater degree of local direction over their economies.

One such project is Valley Woollen Mills in the Codroy Valley of southwestern Newfoundland. Nine villages, comprising 3000 people, joined together to form a Rural Development Association, a format encouraged by the provincial government (Wismer & Pell 1981). In the mid-1970s, the Association, which had already been involved in making improvements in local transportation and health services, became interested in revitalizing the valley's traditional wool industry. They purchased second-hand machinery for the mill and, through the backing of both senior governments, were able to be in operation in 1977. The mill employs ten people and now imports wool from outside the province as well as using local sources. The mill is expected to generate enough surplus to pay for other Association activities.

On the west coast of Canada several communities on Vancouver Island have begun to utilize an alternative 'monetary' system to obtain many of the goods and services they require. It is called LETS, Local Exchange Trading System, and operates like a community bank where members open an account in 'green dollars' to trade locally with other members. The local LETS enrolls members for a modest fee, issues a monthly listing of requests and offers for goods and services, and keeps track of the individual accounts. In a recent listing from the Cowichan Valley LETS there were requests for fruits and vegetables, an extension ladder, dental work, access to riding horses, etc. Offers of furniture making, French tutoring, catered breakfasts, and calligraphy were among the several pages of goods and services available from individuals and business firms. Those with requests wish to spend green dollars for all or part of their needs. Those offering goods and services are willing to take green dollars for all or part of the offer. LETS goes beyond a simple barter system by using the medium of green dollars to facilitate trading among various members, i.e. those paying for a service build up a negative balance and may erase that by providing, say, a service to any member for the same or more green dollars. In its early stages, LETS operates as a manual information system but is designed to operate through an on-line system of personal computers. The Cowichan Valley LETS is organized as a non-profit trust.

DELIVERING SOCIAL SERVICES

The one-on-one contact necessary for dispensing many social services – family assistance, child support, personal counselling, employment assistance, etc. – is difficult to attain in rural areas because of the large distances between clients and service providers. Two rural communities in Ontario tackled the problem of bringing people in their areas social services normally only available in large centres. North Frontenac Community Services, a consortium of half a dozen municipalities in a 2700 km² area with about 5000 people and no settlement over 500 in population, organized a 'clearing-house' for a wide variety of senior government services. Case workers from a variety of programmes are provided with a common base of operations in the largest village. This not only facilitates access to professionals but also allows co-operation amongst professionals when several services may be needed by a client. In another part of rural Ontario, Family Focus, a community non-profit society, serves a three-county area of mostly small towns and farms with social services in eastern Ontario. Its approach is either to obtain funding from the provincial government and deliver services directly or to help consolidate the programmes of existing agencies so that they may be delivered to rural people.

In both these instances, the group has become a focal point for a wide variety of community development activities. The perceptions of local needs are facilitated by their boards of directors drawn from the community. In addition, senior government programmes are able to be delivered more effectively when and where they are needed in these areas. One dilemma is how to extend to other rural areas similar sorts of local effort. However, they require local initiative and cannot be 'force-fed' by government agencies. The form they take and the agenda they propound will vary, as these two cases and many others across Canada attest. Nevertheless, they reflect local needs and capabilities.

MOBILIZING RURAL RESOURCES

The problems and needs of a rural community tend to be discrete: lack of a doctor, or a curling rink, or a restaurant, or new housing, etc. The adjacent community, perhaps, has different needs, or some that overlap with the first. Canadian experience shows that rural people participate in an extended rural community comprising several small centres (Hodge & Qadeer 1983). Given this situation, co-operative problem solving may serve a wider range of needs and attain them more economically. Such an approach is used by a group of ten towns and rural municipalities in Alberta. Regional Resources Project No. 1 was established in 1972 and covers an area of 9000 km² with 13 000 people. It operates through a board of directors, one from each community, and a single staff person who is, in effect, the consortium's 'planner'.

The Project's Board provides a forum wherein the ten communities plan for their various capital projects and services in a co-operative manner (Bodmer 1980). For example, if Acme is planning to build and lease a hardware store, Beiseker, half a dozen miles south, might be advised to invest in another kind of facility. The project co-ordinator provides the link between communities in such situations, as well as helping them to mobilize resources for projects. Several pairs of communities have co-operated to build shared recreation facilities. Some communities have embarked on small housing projects to meet local shortages. Local development corporations are frequently formed to raise locally the necessary start-up capital. The distinctive feature of this undertaking is the commitment to a formal structure to ensure that the oft-spoken ideal of intermunicipal co-operation becomes a reality.

RURAL REGIONAL PLANNING

Regional planning may be established to serve senior government objectives, as was discussed above, or what can be called regional planning 'from the outside'. But regional planning also may be established to serve a region's needs and directed in that quest by regional interests. Many regional planning agencies for rural regions have opted for planning 'from the inside'. Two noteworthy examples are for the Restigouche region of New Brunswick and the Peace River region of Alberta. Both of these agencies have developed plans for community facilities, housing, recreation, transportation, economic development and environmental protection which started with the active participation of the region's residents in identifying goals. Throughout the plan-making and implementation there is co-operation and consultation between the agency, the citizens and their local governments.

Furthermore, the style of planning of these agencies is *positive* rather than *adaptive*. The latter too often seems to be the mould that rural planning falls into because of the dependency relationship. In the case of the Peace River region, where major resource development projects are likely, the Commission asked for a 'seat at the table' with the provincial government when such projects were being considered as well as to conduct local public hearings.

Reprise and prospect

Over the past four decades, there have been many public planning efforts *directed at* rural areas in the name of economic development, conservation of resources, preventing scatteration, and so forth. In most cases the planning was for the achievement of national or provincial objectives. An understanding of local needs was seldom sought. The results of these directive actions range from fiasco to fleeting success. Some rural communities suffered irreparable damage, such as those in Newfoundland. Others gained, perhaps, a new factory, better highway access or a zoning by-law. Survival was offered to rural communities but not community-building. The 'Stay Option' of Manitoba was one of the rare exceptions in this top-down planning.

Rural communities did survive and their populations have been increasing. Nearly 60% of the 9100 towns and villages in Canada had larger populations in 1981 than in 1961. All rural regions experienced this 'staying power' to a greater or lesser degree. That rural communities survived is due, it seems, less to planned large-scale projects and development programmes or the use of urban planning concepts in rural situations than to three mundane reasons. First, there has been the extension of public services (health, education, social welfare) and physical infrastructure (highways, electricity, telephones) to all segments of society, rural as well as urban. Rural areas often saw local hospitals and schools closed to satisfy system-wide standards of efficiency, but the overall outcome has been to allow rural residents to function much as their urban counterparts. The second reason is the out-migration of population from rural to urban areas which helped achieve a more viable balance of population to resources. This was especially true up to the early 1970s. Recent positive net migration to rural areas seems related to the improved living conditions cited above. In any event, substantial numbers still leave rural areas and, like most migration in Canada, is not a product of deliberate programmes to shift populations.

The third reason for rural community persistence is the self-reliance shown by rural residents to improve their own situations. The diversity of these efforts, their small scale and independence from established programmes make it difficult to determine rationally their net contribution to community welfare. Nevertheless, they are widespread and many rural people contribute resources to them. Further, we should note two features of community-based development: one is the wide array of approaches and projects entered into by rural communities and the other is the high degree of planning that is utilized. If top-down planners could appreciate the diversity of local needs and also the willingness of local people to participate in the planning of programmes, there would likely be much less animosity than there has been.

The outlook for Canada's rural areas is that two broad inherent issues will have to be dealt with and two broad external pressures will have to be coped with. The two internal issues are the settlement of claims by indigenous peoples to extensive areas of land and the needs of a very fast increasing elderly population. Each in their own way affect the delivery of senior government programmes and the objectives of natural resource development. The two external pressures will result from the free-trade initiatives of the federal government and from the continued infatuation of senior governments with mega-projects. Free trade, whether fully achieved, will affect the production, the employment and the income associated with natural resources, which come almost entirely from rural regions. Many mega-projects are 'on the drawing boards' – from harnessing tidal power from the Bay of Fundy to transporting oil and gas through the Mackenzie valley – that will have major impacts on their rural regions. At this point, one can only hope that future rural planning will have learned to avoid the pitfalls of the disjointed and insensitive planning of the past.

Notes

1   In this discussion 'rural Canada' includes a broader base of population than that used by Statistics Canada in the national census. The census defines all population concentrations of 1000 or more and with a density of 400 per square kilometre as *urban*. The author believes that conditions of rurality persist in regard to all towns and villages up to a population of 10 000.

2   The Canadian census employs a category called 'Rural Non-farm' that includes, without distinction, both open countryside and hamlet and village (under 1000) population.

References

Baker, H. 1960. The impact of Central Government services on the small community. *Canadian Public Administration* **3**, 97–106.

Berger, T. 1977. *Northern frontier northern homeland*. Toronto: James Lorimer.

Bodmer, H. 1980. Regional Resources Project No. 1: an innovative approach to economic and social development. *Plan Canada* **20** 2 (June), 81–90.

Copes, P. & G. Steed 1975. Regional policy and settlement strategy: constraints and contradictions in Newfoundland's experience. *Regional Studies* **9**, 93–110.

Hodge, G. & M. A. Qadeer 1983. *Towns and villages in Canada: the importance of being unimportant.* Toronto: Butterworths.

Matthews, R. 1976. *There's no better place than here.* Toronto: Peter Martin.

Newfoundland Department of Municipal Affairs 1968. *Planning for smaller communities.* St John's: Project Planning Associates.

Pross, P. 1975. *Planning and development: a case study of two Nova Scotia towns.* Halifax: Dalhousie University Institute of Public Affairs.

Qadeer, M. A. 1979. Issues and approaches in rural community planning in Canada. *Plan Canada* **19** (2), 106–21.

Qadeer, M. A. & K. Chinnery 1986. *Canadian towns and villages: an economic profile,* 1981. Winnipeg: University of Winnipeg Institute of Urban Studies.

Russworm, L. H. & C. R. Bryant 1984. Changing population distributions and rural–urban relationships in Canadian urban fields. In *The pressures of change in rural Canada,* M. Bunce & M. Troughton (eds), 113–37. York University Geographical Monographs, Toronto, No. 14. Toronto: York University.

Sternlieb, G. & W. Hughes 1977. New regional and metropolitan realities for America. *Journal of the American Institute of Planners* **43** (3), 227–40.

Wirth, L. 1938. Urbanism as a way of life. *Urban Sociology* **44**, 8–20.

Wismer, S. & D. Pell 1981. *Community profit.* Toronto: Is Five Press.

9 *Australia*

J. H. HOLMES

In this chapter it is argued that, in Australia, rural issues exercise a powerful influence on federal and state policies, programmes and institutions, disproportionate to population numbers and level of economic activity in the rural sector.

It is also postulated that this strong rural bias, which paradoxically exists in a nation with the highest global levels of urbanization and metropolitan concentration, is nevertheless consistent with widely supported national goals and aspirations. These goals are frequently invoked as the rationale for an impressive army of programmes designed to promote rural development and to ameliorate the perceived disadvantages of living in Australia's thinly populated rural areas.

It can also be shown that this ongoing attention to rural priorities is consistent with the wider historical and geographical context of Australia's economic development. Notwithstanding its status as an advanced Western nation and reasonably high living standards, Australia can still be classed with the hinterland economies, reliant upon exports of raw materials, namely agricultural products and minerals, with a continuing strong comparative advantage in its wealth of natural resources, especially land, relative to its small population. There remains a strong public perception of Australia as being relatively undeveloped and underpopulated, with vast opportunities for further growth based on natural resources. The philosophy of resource development remains entrenched at all levels of government, federal, state and local, and is most evident in the so-called frontier states and territories, notably Queensland, Western Australia and Northern Territory, as so strikingly revealed in Harman & Head (1982).

Yet rural development in Australia has persistently been handicapped by severe environmental and locational constraints (Blainey 1966). Australia did not experience the same 'incessant expansion' or 'remorseless, surging movement of the frontier' as described by Turner in the United States (Allen 1959: 113). The Australian 'frontier' was one of high risk and many failures. This has tended to prolong the frontier phase in Australia's settlement history, with the vast arid interior and the tropical wet–dry northern zone still being regarded as major frontier zones, awaiting the inevitable impetus for development.

The rural development challenge has been further heightened by the seeming paradox of extreme urbanization in a national economy with a predominantly rural export base. In his authoritative work on Australian economic growth in the latter part of the 19th century, Butlin states that 'the process of urbanization is the central feature of Australian history, overshadowing rural economic development and creating a fundamental contrast with the economic development of other "new" countries' (Butlin 1964: 6). Butlin points out that by 1891 two-thirds of the Australian population lived in cities and towns, a fraction matched by the USA only by 1920 and by Canada not until 1950. Urbanization is linked in part to the highly commercialized, export-oriented economy and the high level of employment in the service occupations, but it is also strongly reinforced by the very low labour requirements and high service demands of the rural economy. This is clearly expressed in the contrasting levels of labour productivity. Labour productivity in the agricultural and mining sector are by far the highest in the world, matched only by New Zealand. Australia's strongly export-oriented agricultural and mining sectors employ only 6.2 and 1.4% respectively of the active workforce. Labour productivity in manufacturing and service industries is much lower, and labour demand is focused in these urban-based sectors.

Allied to urbanization is the extreme degree of population concentration with over two-thirds of Australians living within the five core coastal metropolitan regions, centred on Sydney, Melbourne, Brisbane, Adelaide and Perth. This concentration is a logical response to Australia's large size and small population. The heavy burdens of transportation and servicing a dispersed population are minimized, and economies of scale arising from agglomeration can, in part, be realized.

Notwithstanding recent evidence of a modest, locationally selective trend towards counterurbanization, in favoured coastal regions near major cities, these sharp contrasts in urbanization and metropolitan concentration tend to be self-reinforcing. The outcome is a very thin dispersal of the residue of the population over a vast land area, with attendant problems in capturing any local multiplier effects from new resource-based ventures, and with continuing problems of delivering essential services to the scattered population.

Australia's chequered frontier history has not only ensured a continuing emphasis on resource development but has also created a social and political ethos in which governments are called upon to intervene strongly in support of such goals. Whereas traditionally the American frontiersman was hostile to government interference, in Australia the ardent protagonists of development have argued vociferously in favour of government intervention. This tradition became entrenched during the 1860s, as the six self-governing colonies vigorously pursued a rural development programme designed to populate Australia's vast empty spaces. The large sheep 'runs' of the pioneer squatters were to be broken up into smaller 'selections' or family-sized farms. These programmes of closer settlement were to be achieved using the twin instruments of land legislation and infrastructure provision. As in the USA, land laws enabled selectors to take up land for farming. However, unlike the USA, Australian colonial governments also accepted the responsibility of providing needed rural services. This expanded rôle of governments reflects, in part, the greater difficulties of pioneer settlement in Australia. There were many failures; successes depended heavily on persistent governmental support. Small populations, long distances, low demand levels and a limited local tax base all combined to deter private investment or local initiative in the provision of essential services. Costly services could only be sustained by the intervention of the colonial governments who, by default, assumed centralized control of education, police and public safety, hospitals, railways, major roads, and postal services. With federation in 1901, the six states retained most of these responsibilities, transferring only postal and telephone services to nationally controlled public utilities. Although centralized control brought many rigidities, it was the only feasible way to foster rural development at the desired pace. It also ensured the maintenance of reasonable standards in the provision of education and health services, which local initiative could not achieve because of lack of revenue.

Centralized bureaucracies were established not merely to ensure adequate service provision in sparsely settled rural areas but also to foster and guide rural development. Initially, the Lands Departments in the colonies occupied a pre-eminent position, with their very considerable powers in regulating the use of Crown land, held mainly under lease or licence, and with the prime function of ensuring the subdivision of lands into family-sized 'home maintenance areas'. In their own spheres, Departments of Mines, Forestry, Water Supply and Irrigation exerted comparable influence. These bureaucracies have strong vested interests in ensuring that public intervention continues to be strong, and is generally directed towards support of rural interests.

Rural development programmes

Programmes designed to foster rural development and population decentralization have changed appreciably, though often belatedly, in response to changes in economic opportunity. The changing emphasis in major programmes since 1861 is suggested in Figure 9.1. Some of these programmes have a strong regional or locational focus, whereas others have state-wide or nation-wide applicability.

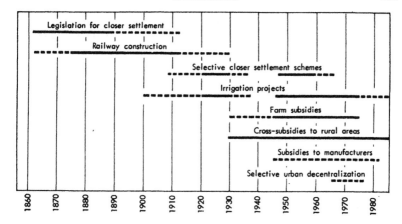

Figure 9.1 Chronology of major governmental programmes to promote rural development and decentralization.
Source: Holmes (1987).

The first significant legislative actions fostering rural development were the New South Wales Robertson Land Acts of 1861, heralding a succession of laws intended to encourage a second wave of pioneers to occupy the vast unimproved grazing 'runs' of the original squatters, thereby establishing closer settlement in suitable areas. Government support came mainly through legislation intended to reduce the costs and remove other barriers to the acquisition of land by smallholders. Major public investment was soon to be channelled into railway construction, proclaimed as the solution to problems of inaccessibility for small farmers in inland Australia. The successes and failures of closer settlement have been well chronicled (Roberts 1924, Allen 1959, Meinig 1962, Powell 1970, Williams 1975). Save in a few areas of belated pioneering, closer settlement has steadily lost momentum during the 20th century. Governments attempted to restore this momentum by instituting programmes of concentrated or selective closer settlement, most evident in the soldier settlement schemes following World Wars I and II. Particularly in the 1920s these schemes experienced extreme difficulty through bad planning, undercapitalization and lack of markets, with such consequent hardship and loss that very few closer settlement schemes outside of irrigation have been since attempted, the most recent being the marginal wheatlands schemes in Western Australia and brigalow clearing and development programmes in central Queensland.

For most of the 20th century, public-financed irrigation schemes have been regarded as the most effective way of increasing rural population. Their success in achieving this specific objective is evident in the spectacular localized increments to rural populations along the Murray River and its tributaries. However, their net effect upon Australian economic development has been the subject of considerable controversy (see Davidson 1969).

Consistent with national optimism about Australia's future, great faith was attached to rural development projects, which were regularly publicized in the media, loomed large in the annual reports to parliaments, and figured prominently in election campaigns both at state and federal level. These projects varied considerably in scale, cost, duration, complexity, immediate purpose and level of success.

Public opinion and political programmes were strongly oriented towards state participation: the appropriate legislative and administrative machinery had been created, and the necessary technical expertise had been acquired; expectations remained high that government investment would yield very favourable returns; the local benefit was seen to be consistent with national welfare, so that 'pork barrelling' became an effective political tactic. Hence, a continuing commitment to national development became a political philosophy, sustained by public and private interest groups at national, state and local level, and with the appropriate instruments at hand. Projects with similar constructional requirements and technical demands gathered their own momentum, most notably programmes for railway construction, dam-building for irrigation or hydro-electricity, rural electrification, and road construction. Two obvious examples have been the Snowy Mountains Commission and the Tasmanian Hydro-Electricity Commission. At times these instrumentalities became powerful agents for regional development, by virtue of their in-house expertise and bargaining power in higher political circles. Resource-development projects were the visible expression of a seemingly coherent national development policy, or, more correctly, one federal and six state policies, all of which were founded upon a single-minded pursuit of growth and dispersal of population and economic activity.

Tariff compensation, balance of payments and agricultural support programmes

Following the federation of the six colonies in 1901, protectionist tariff policies were adopted to foster the undeveloped manufacturing sector. Only in the 1920s did the cost burden imposed on agriculture by protectionist policies become a source of dissension, providing the motivating force behind the creation of the Country Party, which has since maintained an active rôle in formulating national policies in such matters as rural assistance, national development, trade and tariffs. A public committee of inquiry, as early as 1929, argued that stable tariffs would have little adverse effect upon rural production costs or incomes, since they would become negatively capitalized through lowered land values. Nevertheless, the argument of 'tariff compensation' has increasingly been used as a rationale for the various agricultural support programmes which have been undertaken by the federal government from the 1930s onwards in response to pressures from the Country party and from spokesmen for particular agricultural sectors. Governments have been reluctant to endorse the tariff compensation argument, but have given greater emphasis to the importance of agricultural exports to the national economy:

> It has been pointed out by many commentators that the complex agricultural policy that emerged in the post-war period was strongly influenced by concern for agriculture's contribution to the balance-of-payments under relatively fixed exchange rates. This concern provides some sort of rationale for such diverse policies as home price schemes, tax concessions to agriculture, subsidies for export promotion, government involvement in agricultural research and extension services and public investment in land development and the provision of a rural infra-structure. These measures were designed to encourage rural output and exports in the face of a balance-of-payments constraint to economic growth. In terms of the responsiveness of rural output (although probably not in terms of cost effectiveness), the overall policy could be judged a modest success. (Edwards & Watson 1978: 190)

In contrast to earlier closer settlement programmes with a developmental thrust, these newly evolving policies gave greater attention to stabilizing and sustaining established agricultural activities, and were often adopted in an *ad hoc* piecemeal fashion to meet the needs of narrow interests within sectors experiencing short- or long-term difficulty, such as dairying, wheat, sugar or dried fruits. A common goal was stability, which was to be attained by price equalization schemes or marketing guarantees, sometimes sustained by direct subsidies. Where the domestic market could be used to support price levels, this was done, most notably to assist the dairying and sugar industries:

> The home price scheme is the major form of agricultural price support used in Australia. An example of price discrimination, home price schemes exploit differences between the elasticities of demand on domestic and overseas markets and require mechanisms such as tariffs or import prohibitions to maintain separation of markets and allow producers the benefit of higher domestic prices. (Edwards & Watson 1978: 190)

This search for stability and guaranteed return was a predictable response to the uncertainties generated by extreme price fluctuations on world markets.

In the two decades following World War II, governments were able to maintain policies designed to generate a modicum of growth in the agricultural sector, with the varied array of agricultural support programmes being supplemented by a continuing series of capital-intensive closer settlement projects often based on irrigation schemes in which the major capital costs were written off from the outset. Increasing public scrutiny of the more wasteful of these programmes, spurred by the cost-benefit analyses of agricultural economists, notably Davidson (1969), forced closer examination of the government's rôle in the rural economy. This re-examination was triggered in the 1960s by prolonged, acrimonious debate about two major programmes, one an industry support programme and the other an irrigation project, both of which were regarded as conspicuously wasteful since they clearly failed to achieve their publicly stated goals. The Dairy Stabilization Scheme was based upon a substantial annual subsidy combined with a discriminatory pricing scheme in which prices for butter and cheese on the domestic market were usually over 80% higher than prices obtained for overseas sales. The exceptionally high transfer payments received by dairy farms were singularly failing to resolve a persistent low-income problem experienced widely throughout the industry but heavily concentrated in the subtropical zone, where agronomic difficulties had not been fully resolved. Critics were able to argue that much-needed farm adjustment was being impeded at very heavy cost to the taxpayer and the consumer (see Throsby 1972).

The second major target for criticism was the Ord River dam and irrigation scheme, located in the remote Kimberley region in north-west Australia. This was an ambitious scheme intended to demonstrate that intensive agriculture could be economically viable in Australia's sparsely settled north, provided that the usual practice of writing off the investment in capital works was followed. The scheme was favoured by very low costs of water storage, and a major agricultural pilot programme was undertaken. In spite of substantial government investment in providing further assistance to the original landholders, irrigation farming has virtually ceased; production costs were much too high, arising partly from agronomic problems but mainly from the additional cost burdens imposed by the isolated location of the project (see Davidson 1965). The Ord River débâcle hastened the demise of government-funded capital-intensive agricultural development projects, just as the follies of the Dairy Industry Stabilization Scheme hastened a shift from farm-support towards farm-adjustment policies.

Rural adjustment programmes

From the mid-1960s onwards, there has been a gradual move towards rationalization of agricultural policies. This rethinking has been spurred by a number of circumstances, and not solely from the transparent deficiencies of some long-established programmes. The increased agricultural protectionism of the European Community and the loss of Australia's main market when Britain joined the Community forced the federal government to undertake an 'agonizing reappraisal' of agricultural support programmes, particularly in those products with little comparative advantage such as dairying and dried and canned fruit. This was further prompted by the mounting evidence of endemic poverty among many producers in these industries. This re-examination was also assisted by a more favourable trade pattern which prevailed during the 'minerals boom' of the late 1960s and early 1970s:

> After the mid-sixties, the environment changed dramatically. 'Balance-of-payments pessimism', which had influenced both policy advice and government decisions for decades, ceased to be relevant. The mining boom had increased exports substantially and exchange rates could be, and were, varied more readily to achieve external balance. Therefore, many of the *ad hoc* measures taken to increase agricultural production and exports would not now be needed (and, hopefully, not even contemplated) even if the same balance-of-payments situation existed. Nevertheless, it has proved politically difficult to change many policies because agriculture has been experiencing considerable difficulties in recent years with low prices, at various times for wheat, wool, beef, fruit and dairy products. Coupled with the exacerbating effect of domestic inflation there has been substantial pressure on farm incomes. (Edward & Watson 1978: 191).

During the 1980s Australia has again been enmeshed in an endemic balance-of-payments problem, created by low world prices for agricultural and mineral products, and exacerbated by strengthened agricultural protectionism in Australian main markets. However, Australian governments appear to have absorbed the hard lessons gained from excessive protectionism, and recognize that Australia's financial problems would be aggravated if costly agricultural support programmes were pursued in competition with such economic giants as the USA and the European Community. This more cautious approach towards agricultural support schemes and greater attention to agricultural adjustment is an indicator that governments recognize Australia's hinterland status, and that Australian farmers must be 'price-takers', not 'price-makers'. Government assistance is increasingly being directed towards easing the path of adjustment and rationalization of rural production units.

Before 1970, adjustment schemes had been very restricted in scope, and responsive only to localized critical farming problems, such as heavy indebtedness among irrigation farmers during the agricultural depression of the early 1930s requiring state irrigation authorities to write off considerable debt from small 'block-holders'. From 1938 to 1944, the Marginal Wheat Areas Reconstruction Scheme provided federal funds to the states to reconstruct holdings in marginal wheat areas, devastated by the drought of the early 1930s.

The only other early scheme concerned with structural adjustment was implemented in New South Wales in 1960. The Closer Settlement Advisory Board, whose established function was to advise on subdividing large holdings for closer settlement, was called upon to reverse its rôle and was enabled to purchase non-viable dairy farms for allotment to neighbours. Up to 1970, when the national schemes became operational, about 280 farms were acquired by the Board.

McKay (1967) presented the first substantive evidence of a widespread farm adjustment problem, affecting all agricultural sectors. Using the surveys of the Bureau of Agricultural Economics, of which he was director, McKay suggested that as many as one-third of Australian farms had incomes below A$2000, and therefore could not operate efficiently because of an inadequate income. By the late 1960s the competitive position of Australian agriculture was such that policies clearly had to be redirected from promoting higher output towards industry rationalization and improved farm productivity. The Marginal Dairy Farms Reconstruction Scheme was implemented in 1970 with funds provided by the Commonwealth government to state reconstruction authorities. The Scheme provided for the reconstruction authorities to purchase marginal dairy farms offered for sale and to make this land available for amalgamation with other land so that the resultant unit constituted an economic holding, not necessarily in dairying. Half of the funds was provided to cover the 'write-off of redundant farm assets' and any drop in the value of land sold in amalgamation. The Scheme thus provided for a higher farm price for the vendor and a lower price for the purchaser than the free market would provide. In 1974 the provisions of the Scheme were extended when the Dairy Adjustment programme replaced the original scheme. Important extensions were the provision of finance for property development and relocation assistance.

The producers of other commodities were also in economic difficulties. Wool prices had declined sharply from the mid-1960s, and by 1971 a large proportion of wool producers earned incomes considered no longer viable.

Recovery prospects were poor with the price of wool apparently set by competition with synthetics. Quotas on deliveries had been applied to wheat producers. Government responded to a depressed rural situation and outlook with a second scheme implemented in 1971 called the Rural Reconstruction Scheme, which provided funds for debt reconstruction, farm build-up and farmer rehabilitation. Farmers producing any commodity were eligible. The 'debt reconstruction' provisions provided funds to farmers unable to meet their current commitments but considered to have sound prospects of long-term viability. State reconstruction authorities, usually the same as those administering the concurrent dairy schemes, could reschedule eligible farmers' current debts and replace some with a long-term debt to an authority. Build-up loans were provided to encourage the amalgamation of holdings considered too small to be economically viable. As with the dairy adjustment scheme, funds were also provided to reconstruction authorities to offset losses sustained in the disposal of 'redundant' assets. However, very little was used for this purpose. Rehabilitation loans were in fact grants to assist farmers who were refused debt reconstruction assistance or who disposed of non-viable units to the authorities and who were leaving farming under severe personal hardship. Use of this measure has been negligible.

Other measures were adopted by government after 1970 to encourage the amalgamation of smallholdings. For example, beginning in 1971–2 the Commonwealth Development Bank extended the Bank's functions to include lending for farm build-up and debt rescheduling. Modifications were subsequently made to these various schemes. A new scheme, the Rural Adjustment Scheme, consolidating the earlier ones into one comprehensive arrangement and introducing further extensions, came into operation in January 1977.

Of the A$228 million expended through the Rural Reconstruction Scheme to December 1976, 46% had been used to finance farm amalgamation. Thus, the bulk of the assistance has gone to debt re-financing which, because it applied only to the operators of holdings judged to be long-term viable units, did not influence the structure of Australian farming or the size of the workforce. Debt re-financing reduced or forestalled capital losses of creditors and the eligible farm operators who in principle would otherwise have been forced to sell up and be replaced by others. Arguments that debt reconstruction measures have promoted resource-use efficiency in agriculture are usually based on the premise that widespread failure of farm business would create instability in farm asset values and exaggerate pessimism, leading to excessive transfer of resources out of agriculture. Reaction to such arguments demands evidence that expectations would become unstable or biased and points to the relatively small number of businesses benefiting from the measure. The numbers of farms receiving direct assistance from the various schemes have been small even, apparently, in relation to the numbers of farms making adjustments. For example, only 80 out of 3244 registered dairy farmers leaving dairying in New South Wales between 1970 and 1974 participated in an adjustment scheme.

Government financial commitments to agricultural adjustment have been relatively modest, not merely because of the limited public funding available but also because Australian agriculture has shown a strong capacity to survive through periods of adversity while also accommodating to long-term changes. Thus the cost–price squeeze, which has been enforced with increasing severity on Australian farms, has been the main impetus for a steady process of farm consolidation such that the number of commercial farms in Australia has declined from 220 000 in 1955 to 174 000 in 1984, and the farm workforce has declined even more dramatically from 460 000 in 1955 to less than 250 000 in 1986. However, employment growth in ser-vices to agriculture has almost balanced the decline in farm employment, such that combined employment has remained fairly constant at just over 400 000 since 1971.

It would be misleading to suggest that policies are now directed solely towards agricultural adjustment, together with a continuing commitment to established price equalization and stabilization programmes. Govern-ments still come under heavy pressure, particularly near election time, to revert to old-style publicly funded special assistance or project develop-ment programmes, particularly in areas of presumed opportunity, such as the north. During 1985–6, the two most striking examples were to be seen along Queensland's sugar coast. Disastrously low world prices for sugar, partly as a consequence of the heavy subsidization of sugar in the European Community, created a severe crisis in Australia's sugar industry, chiefly dependent upon export markets. Confronted by heavy pressure from sugar growers and the Queensland government, the federal government has had to modify its initial proposals for industry deregulation and rationalization, and has agreed to support a minimum producer price, though admittedly at such a low level that the recent modest rise in world prices may yet pre-clude any direct subsidy payments.

Yet, in response to electoral pressures, the federal government is financ-ing the construction of the large Burdekin Dam, near Townsville, with most of its capacity being committed to further expansion of irrigated agriculture. At present it is not possible to identify any crop which is not already oversupplied, so the Burdekin Dam may yet join the Ord as a monument to grandiose notions of northern development.

Australia's northern zone with a tropical wet–dry climate still remains a relatively underdeveloped zone, with some evident potential for agricul-ture. Since mid-century this frontier region has attracted a succession of grandiose agricultural development projects, both publicly and privately funded, all characterized by insufficient initial production trials; by a failure to identify underlying problems and to plan effectively for all aspects of production, storage, transport and marketing; by haste in initiation and by great haste in termination, usually accompanied by heavy financial loss. There also is a strong propensity to attribute failure to some incidental circumstance rather than to assess the underlying problems of initiating complex, highly capitalized schemes in a region lacking essential infrastructure and experiencing extreme locational disadvantage.

Another public policy area of increasing controversy are the drought assistance programmes administered by federal and state governments, including freight concessions on in-movement of fodder and out-movement of livestock, as well as special-purpose low-interest loans. Although there may be net benefits in ensuring the survival of breeding stock in the most severe droughts, there is mounting evidence that current relief programmes are contributing to overgrazing and mismanagement of fragile semi-arid pastoral lands. Relief measures are encouraging landholders to stock heavily and to call for special relief more frequently, as 'dry spells' become increasingly classed as 'droughts' requiring public assistance (see Lovett 1973).

Thus, Australian governments still have difficulty in avoiding the traditional piecemeal, *ad hoc* approach to solving rural problems. This fragmented approach has been reinforced by the division of responsibilities between three levels of government – federal, state and local – encouraging each tier to take credit for any identifiable benefits but blaming others for equally identifiable, unsolved problems. Co-ordination between federal and state governments has been marred by intergovernmental rivalries, by the jealous preservation of 'states' rights' and by the impediments created by the Australian constitution.

Sectionalism has been further promoted by the fragmentation of rural demands by place and by commodity interests. Parochial policies have focused on providing additional benefits to particular rural communities or to organized commodity interests. This mode of representation is so strong that one state Minister for Agriculture made the following comment:

> Farmer organizations are structured on a commodity basis, and carry out their lobbying in that way. For this reason such matters as rural adjustment or improved training for farmers receive scant attention. Farmers see political representation as a means by which they achieve support for the production of their commodities. Their belief in the fetish of commodities is nowhere more apparent than in their political consciousness. They see no inconsistency in biased rural representation, after all, this is more power to the important commodity . . . In the past this philosophy

has been extremely successful and commodity production has been supported long after markets have gone or changed. (Chatterton *in* Van Dugteren 1978: 127).

This 'commodity fetish' has inhibited the development of integrated, flexible agricultural policies. Equally damaging to long-term rural interests have been the enduring disagreements between organizations representing farmers and graziers, respectively. Grazier organizations traditionally have favoured minimal governmental interference with free market processes, recognizing that the greatest benefits to the grazing industries lay in minimizing production costs and maximizing export opportunities. Free-trade policies would best promote these goals. This was an implicit acceptance of Australia's hinterland status, and was intended to maximize reciprocal trade relations with heartland economies. On the other hand, farmers have been more interested in the opportunities presented by the growth of the Australian domestic market, and have accepted, with some reluctance, the need for protectionist policies, provided compensatory assistance and other forms of public support were given to rural industries and communities. Hence the frequent jibes about Country Party politicians being selective rural socialists seeking to socialize the losses and privatize the gains. Fundamental disagreements between the anti-interventionist views of the graziers and the strongly pro-interventionist views of many farmers' organizations have still not been entirely laid to rest even within the recently created National Farmers' Federation, which now acts as a single umbrella organization designed to strengthen the farmers' voice in national affairs. Generally, the farmers' views have prevailed, as governments have responded with programmes tailored to meet specific demands.

A piecemeal approach may have been justified in the expansionist phase of rural economic growth, when a rapid response to new opportunities and demands could be achieved. Such is no longer the case, and political parties are gradually adopting a more realistic package of co-ordinated farm support programmes, realizing that Australian farmers must fend for themselves in an increasingly difficult world market.

Service provision in rural areas

In examining the rôle of governments in rural Australia, it is not enough to consider only agricultural programmes. Governments at all levels, federal, state and local, have many responsibilities in providing an infrastructure of services, within which questions of service quality, accessibility and price loom large. Although policy issues, mechanisms and outcomes are broadly similar in all western societies, important differences can be discerned over time and between countries. In Australia, the task of servicing rural communities and of ensuring equitable treatment for rural populations has been made more difficult by the recency of settlement, by the insistent demands generated by a population growth and spread linked with expectations of further growth, by the vast areas to be served and the extreme sparsity of the population, and by divisions existing between the three levels of government. Commonly, rural development programmes have been intended to serve the same national goals as agricultural programmes, accordingly following along parallel lines, usually requiring a high level of centralized funding and control in the absence of adequate local resources and expertise. This control, however, has mainly been exercised by the six states, either directly or through their tight control over local government.

During the period of frontier expansion and closer settlement, the task of providing the basic rural infrastructure was primarily the responsibility of the six colonial governments, who delegated only limited responsibilities and revenue sources to rural local government instrumentalities, which developed only belatedly in the final decades of the 19th century. Railways, schools, hospitals, post offices and land administration offices were important contributors to rural advancement.

With functions such as education, health, welfare, and public housing being controlled generally by the state governments, and with public utilities such as electricity and water supply being administered by *ad hoc* regional authorities, rural local government has served only a limited rôle, with road maintenance sometimes being the predominant function, particularly in the non-urbanized shires. Local initiatives are further constrained by the limited revenue base of rural local government, based on property rates supplemented by special-purpose grants from state governments, and by the incessant demands imposed by road maintenance within shires which are large in area but small in population and income. Accordingly, rural shire councillors may often consider that they have a more important rôle as local spokesmen seeking assistance from state and federal governments than as decision-makers responsible for choosing between alternatives in funding sources, levels of expenditure and major budget outlays.

This lack of local initiative in service provision has enabled state governments to maintain a decisive rôle in local service provision, although, as with agricultural policy, this rôle has been increasingly challenged by the federal government, which at times has used its powers under the Uniform Tax Agreement to make tied grants to the states, often with strings attached concerning the way in which funds could be used. One notorious case was the requirement that, of the road construction grants reimbursed to the states from petrol tax collections, 40% of funds should be spent on local rural roads. In due course this formula created many anomalous situations with well constructed but little used local roads feeding into badly constructed but more heavily used main roads. This is one of the most obvious of many examples of centralized decision making being highly responsive to strong rural pressures, exercised through the former Country Party, but being incapable of reaching an optimal solution to meet local needs. The issue of costly but inappropriate solutions to rural problems is becoming more acute as rural interest groups argue their case for special assistance on grounds of social justice rather than on the basis of furthering national development objectives.

Equity goals instead of development goals

With agriculture no longer offering major development prospects for the foreseeable future, and with its reduced significance to the national economy, rural spokesmen can no longer argue with much force that special rural assistance programmes are essential for national well-being. Accordingly, increasing emphasis has been placed upon the desirability of maintaining rural communities, ensuring that depopulation and abandonment do not proceed apace and that the hardships of rural living should be minimized. In its most fundamentalist form, presented mainly by spokesmen from the most remote locations, it is argued that all Australians, irrespective of their location, are entitled to equal treatment at equal personal cost in receiving all goods and services, irrespective of the real cost of providing such services.

A more widely used and more generally accepted argument is one which fits closely with the economist's concept of 'merit goods'. It is generally accepted that the basic goals of any advanced, civilized society can only be achieved by guaranteeing all people a minimal level of access to basic requirements as a natural right. Among the more widely accepted merit goods in a Western society are compulsory free education for a minimum period of at least nine years, access to basic medical and health care services and freedom from hunger and extreme want. Generally, the merit goods are those necessary to ensure the development and sustenance of each individual's capacity to fulfil his or her rôle as a responsible citizen within society. Society's responsibility must be extended to cover all people, irrespective of their geographical location.

Australian society has been at the forefront in espousing these goals, and has been particularly responsive to the needs of rural populations, especially in sparsely settled areas. The small one-teacher school, the itinerant teacher, correspondence schools, the School of the Air and a variety of other innovations demonstrate society's concern to ensure that the basic educational needs of isolated children are met. More recently, a battery of financial assistance programmes has been developed to enable isolated children to obtain regular schooling at minimal cost to their families. Although these special programmes for isolated children naturally receive the greatest public attention, it must be pointed out that, *in toto*, there is a much greater investment of material and human resources in ensuring that the variety and quality of education generally available in rural communities and small towns is broadly comparable with that provided in the cities.

Comparable efforts have been made to provide preventive and emergency health services. Although the Royal Flying Doctor Service and Aerial Ambulance Services may be the most spectacular examples, providing an unusually high standard of health care to extremely isolated populations, again a much greater total effort has been expended in ensuring high-quality medical and hospital care generally in rural communities, and in providing speedy referral and transport to specialist facilities in larger centres.

Rural groups have mounted persistent campaigns in favour of expanding the range of services which should be provided 'as of right', and have received sympathetic attention from governments. Because of the challenges posed by distance, sparse population and low levels of consumer demand, in Australia there is a clear relationship between rural population density and the quality and frequency of basic rural services. This has been explored systematically by Holmes (1984, 1987). In the 'closely' settled agricultural areas, rural populations have generally been successful in obtaining all-weather surfaced roads, automatic telephones, mail twice or three times weekly, electricity and daily school access for their children. In the more sparsely settled grazing zone, where extreme difficulties exist in service provision, there has been a prolonged campaign to improve transport and communication services, based largely on the argument that Australians are entitled to ready social access, no matter where they live.

Telecom Australia, in particular, has been subject to strong political pressures to provide services in isolated areas of the same quality and at the same price as those provided to city dwellers. In response to these pressures, Telecom Australia in 1979 announced a Community Access Programme in which one major component was the provision of all rural subscribers with access at local call rates to their nearest major service town; in some cases this nearest town is up to 300 km away.

Major instruments in support of servicing the rural areas are the centralized public monopolies controlling railways, mails, telephones and electricity; other public services such as education, hospital funding and police are centrally administered for similar reasons. Certain services in the private sector may receive direct subsidies in favour of price equalization, as with fuel and domestic liquid-petroleum-gas, or to ensure adequate service, as with some inland air services. Also there are special tax concessions for residents in a defined remote zone, closely coincident with the pastoral zone.

This elaborate array of mechanisms has been used to ensure that public investment in the pastoral zone is maintained or increased; but private investment is in decline. Railway services have been maintained, even where traffic is pitifully sparse. Although the Beef Roads Scheme has ended, there is a continuing programme of special funding for feeder roads, and generous support is provided to pastoral shires for continuing road improvement works. Since 1979, Australia Post has been pursuing a nationwide programme of upgrading once-weekly mail services to a twice-weekly frequency. New mail services will be commenced where delivery costs are no higher than 20 times average urban costs. In 'special circumstances', which are found to exist in almost every case, new services can commence where delivery costs are 40 times higher.

Telecom Australia's ambitious Rural and Remote Areas Programme has the objective of replacing the obsolete PPE (Part-Privately Erected), operator-connected, party-wire system with a microwave relay service using digital radio concentrators. In remote areas installation costs will exceed A\$ 30 000 per customer, yet Telecom Australia is under strong political pressure to reduce its connection fees, which at present have a maximum value equal to a capital charge of A\$ 1750 per subscriber.

The domestic communication satellites, estimated in May 1982 to cost A\$ 650 million, have been justified mainly on the grounds of providing improved services to remote consumers; they will provide complete nationwide radio and television relays while also enhancing many other services, notably the School of the Air.

Rural electrification schemes are being extended to smallholdings in the pastoral zone, into areas where up to 30 km of line are required per consumer. In Queensland's Central West district, the rural development programme for the years 1985–7 connects 327 new customers at an average capital cost of over A\$ 40 000 per customer. At one stage the Queensland government was studying the feasibility of installing separate publicly funded diesel generators in remote homesteads as an adjunct of the public electricity supply system, but this has been found to involve prohibitively high maintenance costs (Garland 1983).

Correspondence education has been enhanced and given a more strongly personal, local orientation, with the introduction of activity days, group camps, tours, school visits, training programmes for parent-supervisors, twice-yearly visits by itinerant teachers and some efforts to decentralize correspondence teaching centres. Also videocassettes have been introduced, two-way radio lessons improved, and further upgrading will be achieved using satellite-relayed lessons.

Isolated children in the pastoral zone still usually travel tremendous distances to boarding schools for secondary education in large towns, and little effort has been made to improve local high school facilities. However, financial assistance has been progressively increased in support of travel, boarding and tuition costs for isolated children (Tomlinson & Tannock 1982).

Only modest improvements have recently been made in rural health services, which, in any case, have been at a very high standard, given the sparsity of population. Remote outback locations serviced by the Royal Flying Doctor Service have a higher-quality medical service, for both preventive and emergency treatment, than do some outer suburbs, and certainly a better service than most of the smallholders dispersed throughout the eastern pastoral zone who must rely on medical services in the local town which may be 150 or more kilometres away along a difficult, sometimes impassable, road. For many remote residents, the most important recent improvement has been the provision of special financial assistance to cover travel and living costs for periods spent away from home receiving specialist medical treatment.

Increasing public investment in rural services is having a powerful impact, not only in enhancing accessibility and living conditions for rural populations but also in generating economic activity and employment in country towns. Public sector activities are becoming increasingly vital to these towns, particularly in the most thinly settled areas, where they help to counter the persistent downward trend in private sector employment. One further major advantage is that the public sector is immune to the vagaries of the local rural economy, which notoriously leads to 'boom-and-bust' conditions for farm-based country-town enterprises. Indeed, public sector activities can be purposefully expanded when the rural economy is in recession. This brings a stronger element of stability into the economies of small towns, and has a powerful influence in ensuring their survival (Holmes in press).

Social justice issues

The increasingly frequent invocation of social justice goals is but one indicator of major long-term changes in the direction and intent of rural policies in Australia, in response to changing circumstances as well as changing aspirations and demands of rural populations. The reduced emphasis given to goals of rural development in favour of goals of social justice is only a further expression of earlier basic shifts away from sponsorship of closer settlement towards assistance with farm adjustment, away from farm sector support programmes towards broadly focused rural assistance programmes, and away from productivity growth towards infrastructure provision.

More broadly, it represents a transfer of formerly urban-oriented social policies into the rural context, most notably policies relating to service accessibility and social justice. Rural populations are no longer asking to be treated differently from urban populations; rather they are arguing for the same treatment, demanding equality in access to services, of the same quality and at the price as available to city dwellers. The major thrust of the argument is that locationally disadvantaged groups should receive equitable treatment in the same way as socially disadvantaged groups. It is argued that the hardships of rural life are already enough without being further burdened by the heavy additional costs in obtaining basic services (see, e.g., Van Dugteren 1978). This argument can be questioned on grounds of equity and efficiency.

EQUITY ISSUES AND RURAL DEPRIVATION

Equity arguments cannot readily be transferred from the socially disadvantaged to the locationally disadvantaged. Many of the latter can well afford to bear their own servicing costs and scarcely deserve costly special public assistance in preference to more needy groups. Indeed, even in rural and remote areas, the emphasis upon basic utilities can be highly discriminatory, providing special benefits to property-owners and overlooking those in greater need.

National attention was first drawn to the overall needs of Australia's poor through the detailed reports of the Commission of Inquiry into Poverty (1975, 1977), which also highlighted the incidence of rural poverty and the extreme problems posed in providing effective delivery of health, education, welfare and social services in most rural locations, and particularly in the most remote areas.

Walmsley's (1980) study provides further confirmation that there is a significant problem of social deprivation in rural Australia, and that the incidence is most severe in remote areas where assistance is least likely to be available. Figure 9.2 shows a close relationship between remoteness and reduced life chances. Walmsley's index of life chances included indicators on employment (5), education (3), social stability (2), social belonging and participation (2), living conditions (5) and family status (3), all being derived from 1976 census data. The most deprived groups include aborigines, itinerant workers and also less skilled workers and their families scattered in small roadside and railside settlements across Australia's vast interior.

Although aborigines are belatedly receiving very special attention in

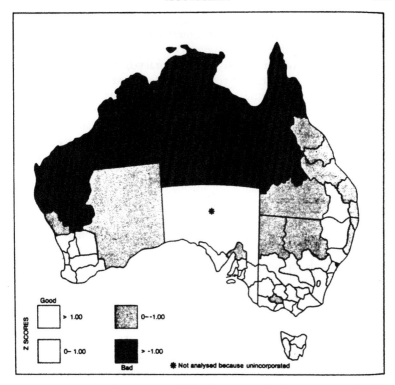

Figure 9.2 Regional variations in life chances in Australia, using weighted 3-scores for 20 indicator variables.
  Source: Walmsley (1980).

overcoming their very severe educational, health and other social disabilities, generally these programmes are ill-suited to their needs, and all the symptoms of extreme social deprivation are clearly evident (Young 1981, Loveday 1982). Non-aboriginal rural groups experiencing social deprivation are still largely overlooked, with their problems remaining either unrecorded or too difficult to resolve because of their scattering over vast distances. Specialist social welfare services have yet to be developed to match the other costly rural services already described.

EFFICIENCY ISSUES AND THE RÔLE OF THE PUBLIC SECTOR

The efficiency counter-argument also carries considerable weight. Normally, there are sound economic reasons for the locationally disadvantaged to choose to live where they do. Indeed, in rural Australia higher incomes are usually a compensatory benefit for populations living in more isolated locations, enabling the population to adjust their livelihoods and lifestyles to their remote locations. An efficient rural settlement and land-use pattern is one which is capable of sustaining its own costs for infrastructure provision. Unduly heavy transfer payments to rural populations will serve only to promote inefficiencies and retard any needed adjustment by the less viable enterprises. Because of the relative smallness of the Australian economy and the substantial dimensions of the task of providing a full array of services to the rural population, including the most isolated groups, any sustained effort to upgrade a wide range of rural services to city standards and at city prices to the users may ultimately impose very heavy strains on the national economy.

Public policies of guaranteeing adequate services in unprofitable rural locations also have a strong influence on such major public controversies as proposals for the privatization of public service monopolies, notably Australia Post and Telecom Australia. Privatization is strongly supported by the free-marketeers of the New Right and by many economists, but is equally strongly opposed by an 'unholy alliance' of conservative rural 'socialists' within the National (formerly Country) Party, the entrenched public service trade unions and the Labour Party. The main thrust of the argument against privatization is clearly stated in a recent news report:

> In a speech delivered for him at Kunnunurra in Western Australia by the Minister for Finance, Senator Walsh, Mr Hawke urged delegates to a Northern Australia Development Conference to 'contemplate what privatisation would imply'. The Prime Minister said a profit-oriented private company would not tolerate services being provided below cost price. 'But setting telephone and postal services high enough to cover costs in country areas would choke off most of the demand, thereby lifting unit costs to even higher levels. The pursuit of cost recovery would be a chase after a vanishing target. What would vanish would be most phones and mail services outside Australia's major cities.' (*The Australian* 1986)

These policy issues have been further reviewed elsewhere (see Holmes 1981, 1985, 1987).

Rural Planning

In contrast with traditional national preoccupations with rural development, rural service delivery and publicly funded resource projects, Australian governments have not revealed much interest in integrated regional or rural planning, nor in exercising strong controls over rural land use. This lack of involvement in comprehensive rural planning is a reflection in part of lack of any urgent need for such planning, combined with an institutional framework which is ill-suited to such initiatives.

As shown earlier, public intervention in rural Australia has been strongly focused on goals of economic development, involving a partnership between public and private capital. The interventionist rôle of governments became progressively reduced as developed goals were achieved. In closely settled areas a *laissez-faire* attitude to rural land use on private landholdings has evolved, with a reliance on marketplace forces. With a relatively plentiful supply of land and with little competition, there was little need to regulate land use.

Integrated rural planning was also impeded by the high degree of centralization and sectoralization of public bureaucracies, each with clearly defined responsibilities, with a strong vertical organizational structure and with little interdepartmental co-ordination. This fragmentation of rural 'planning' between entrenched public departments was further reinforced by the lowly status of local government which possessed neither the powers, nor the resources, nor the skills, nor the will to engage in effective, comprehensive local planning (Logan 1982, Kelleher 1982).

This absence of planning was of no great consequence while there was little pressure to convert land from agricultural and kindred purposes, or any overt conflict in the demand for land, or any public concern on issues of environmental management and land use. Only since the mid-century have such issues emerged, and these have been largely confined to the immediate hinterlands of the major cities, together with prime coastal zones near main population centres. Thus the impetus for land-use controls has arisen from basically urban demands on prime rural lands in the vicinity of major cities, with the most noteworthy legislative responses being in New South Wales, Victoria and South Australia.

Thus the New South Wales State Planning Authority Act 1964, which empowered local shires to extend planning controls beyond urban areas, was nevertheless primarily concerned with settlements. Rural lands were generally classed as 'non-urban', with no attempt to designate appropriate uses, and the 1964 Act specifically prevented planning schemes from regulating agriculture. Prime planning concerns in rural areas were to regulate ribbon development on major roads and to place limits on speculative residential subdivision, particularly in coastal locations.

These initial legislative responses soon proved inadequate in the face of growing pressures on rural lands near major population centres. The most serious planning problems arose from the rapid expansion of rural residential subdivisions from the mid-1960s onwards, with very large areas of former farming, grazing and bushland being subdivided into lots, commonly varying in size from 2 to 40 ha. Even near cities there is a plentiful supply of relatively cheap land available for such purposes, with many less affluent farmers sorely tempted to obtain windfall gains by subdividing all or part of their farms, for a swelling tide of speculators, rural retreaters, hobby farmers, weekenders, retirees and commuters. The initial eagerness of shire councils to encourage such subdivisions, in order to increase their population and tax base, was soon superseded by growing caution as serious problems appeared in the less successful subdivisions, with excessively high land values, low rates of full-time occupancy and use, land mismanagement, poor accessibility and inadequacies in servicing, as well as growing conflicts between different classes of landholders. In the areas subject to pressure for rural residential subdivisions, shire councils now exercise control by zoning regulations setting minimum sizes for any rural land subdivisions, or expressly prohibiting such developments.

Rural residential development is one of the most conspicuous expressions of the general counterurbanization trend which first became evident in the 1976–81 intercensal period. There was a reversal of the long-term trend towards metropolitan concentration of population, with cities above 100 000 persons experiencing below-average growth rates, and with smaller towns and the dispersed rural population growing at rates well above the national average (see Hugo & Smailes 1983). However, in Australia, counterurbanization is a locationally selective trend, as is clearly evidenced when population change within settlement zones is examined.

Goddard (1983) has divided Australia into five distinct settlement zones, of which only one experienced rapid inwards migration over the intercensal period. This is Zone IBI, the closely settled, environmentally attractive coastal zone of New South Wales and Queensland, with its high accessibility and warm subtropical climate, providing a 'sunbelt' image comparable to that in the USA (see Fig. 9.3.). This zone achieved an average migration gain of 2.5% per annum over the period, whereas the other four zones recorded negligible overall trends from migration. However, it is worth noting that Zone IA, metropolitan areas, recorded the second highest average migration gain of 0.4% per annum, with growth in the outer rural areas outstripping decline in the cities.

Counterurbanization is only one component of growing urban-originating pressures on the more favoured and more accessible rural lands, involving not only residential pressures, both urban and rural, but also a rapid growth in demand for open-space recreation activities as well as for other urban purposes, including water supply, quarries and transport corridors. These pressures are occurring in a period of growing environmental awareness, with pressure groups seeking to minimize loss of amenity and also to safeguard prized natural areas, with a strong focus on remaining pristine wilderness lands, over which Australia should exercise a

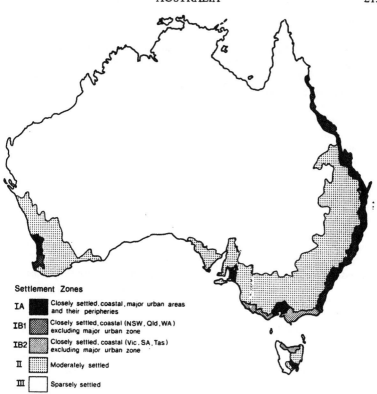

Figure 9.3 Settlement zones within Australia.
Source: Goddard (1983).

responsible custodial rôle because of the rarity of such lands, globally. Environmental awareness also extends to the demand for effective management of Australia's land and water resources, requiring stronger public intervention on the previously unfettered rights of private landholders. There has been a flurry of policy formulation and legislation in these new areas of public concern, with frequent conflicts between the three levels of government – federal, state and local. This represents a major refocusing of rural policy directions away from traditional goals of rural development and resource exploitation and towards new goals of social justice, amenity and environmental protection. It also involves a redirection of public concern away from the more remote frontiers of rural settlement and towards the population core regions and adjacent long-settled rural lands.

References

Allen, H. C. 1959. *Bush and backwoods: a comparison of the frontier in Australia and the United States.* Lansing: Michigan State University Press.

Blainey, G. 1966. *The tyranny of distance: how distance shaped Australia's history.* Melbourne: Sun Books.

Butlin, N. G. 1964. *Investment in Australian economic development.* Cambridge: Cambridge University Press.

Commission of Inquiry into Poverty 1975. *First main report: poverty in Australia.* Canberra: Australian Government Publishing Service.

Commission of Inquiry into Poverty 1977. *The delivery of welfare services.* Canberra: Australian Government Publishing Services.

Davidson, B. R. 1965. *The northern myth.* Melbourne: Melbourne University Press.

Davidson, B. R. 1969. *Australia wet or dry?* Melbourne: Melbourne University Press.

Edwards, G. W. & A. S. Watson 1978. Agricultural policy. In *Surveys of Australian economics*, F. H. Gruen (ed.). Sydney: Allen & Unwin, 189–239.

Garland, K. D. 1983. Rural electricity in Queensland. In *Proceedings of Remote Area Power Supply Workshop.* National Energy Research, Development and Demonstration Programme, Department of Resources and Energy, Canberra, 36–52.

Goddard, R. F. 1983. *Rural renaissance – but where?* ANZAAS Congress, Perth. Conference paper.

Harman, E. J. & B. W. Head (eds) 1982. *State, capital and resources in the north and west of Australia.* Perth: University of Western Australia Press.

Holmes, J. H. 1981. Sparsely populated regions of Australia. In *Settlement systems in sparsely populated regions: the United States and Australia*, R. E. Lonsdale & J. H. Holmes (eds). New York: Pergamon.

Holmes, J. H. 1984. Australia: the dilemma of sparse population and high expectation. In *Rural public services: international comparisons*, R. E. Lonsdale & G. Enyedi (eds). Boulder, Colo.: Westview Press.

Holmes, J. H. 1985. Policy issues concerning rural settlement in Australia's pastoral zone. *Australian Geographical Studies* **23**, 3–27.

Holmes, J. H. 1987. Population. *In Australia: a geography*, Vol. 2, Space and Society. 2nd edn. D. N. Jeans (ed.). Sydney: Sydney University Press, 24–48.

Holmes, J. H. (in press). Private disinvestment and public investment in Australia's Pastoral Zone: policy issues. *Geoforum.*

Hugo, G. J. & P. J. Smailes 1983. Urban–rural migration in Australia: a process view of the turnaround. *Journal of Rural Studies* **1** (1), 11–30.

Kelleher, L. 1982. Aspects of planning in rural Victoria. *Law Institute Journal* **56** (11), 919–28.

Logan, M. I. 1982. Planning and conservation in Victoria: an overview. *Law Institute Journal* **56** (11), 890–4.

Loveday, P. (ed.) 1982. *Service delivery to remote communities.* Darwin: North Australia Research Unit.

Lovett, J. V. (ed.) 1973. *The environmental, economic and social significance of drought.* Sydney: Angus & Robertson.

McKay, D. H. 1967. The small farm problem in Australia. *Australian Journal of Agricultural Economics* **11**, 115–32.

Meinig, D. W. 1962. *On the margins of the good earth: the South Australian wheat frontier, 1869–1884.* Chicago: Rand McNally.

Powell, J. M. 1970. *The public lands of Australia Felix: settlement and land appraisal in Victoria, 1834–91.* Melbourne: Oxford University Press.

Roberts, S. H. 1924. *History of Australian land settlement.* Melbourne: Macmillan.

Throsby, C. D. (ed.) 1972. *Agricultural policy.* Melbourne: Penguin.

Tomlinson, D. G. & P. D. Tannock 1982. *Review of Assistance for Isolated Children Scheme.* Canberra: Commonwealth Department of Education.

Van Dugteren, T. (ed.) 1978. *Rural Australia: the other nation.* Sydney: Hodder & Stoughton.

Walmsley, D. J. 1980. *Social justice and Australian federalism: an enquiry into territorial justice and life chances in Australia.* Armidale: Department of Geography, University of New England.

Williams, M. 1975. More and smaller is better: Australian rural settlement 1788–1914. In *Australian space Australian time – geographical perspectives*, J. M. Powell & M. Williams (eds). Melbourne: Oxford University Press.

Young, E. 1981. *Tribal communities in remote areas.* Canberra: Australian National University.

10 New Zealand

RICHARD WILLIS

Geographical and economic background

Any discussion of rural policies and plans in New Zealand must be set against a rather unique geographical and economic background. A remarkably small population of slightly more than 3 million people (less than the city of Sydney) inhabits two main islands with a land area larger than the United Kingdom, or larger than the combined areas of Denmark, Switzerland, Austria, the Netherlands and Belgium. This population is supported by an economic basis historically dominated by farming, which provides over 60% of foreign exchange earnings (1986) but only 7% of gross domestic product (1984) and employs only about 10% of the labour force (1981).

'It is perhaps predictable therefore that rural studies in New Zealand are strong in their orientation towards agriculture' (Cloke 1986: 3). The historical economic importance of agricultural activities has ensured that in New Zealand rural policies have been synonymous with farming policies; successive governments have sought to preserve and promote the health of New Zealand's most important export industry, at the same time keeping a wary eye on the political climate of the rural electorates which often put them in power. Even in 1986, in a Parliament dominated by a Labour government which has drawn its support almost exclusively from urban constituencies, there are 25 members out of a total of 95 who are farmers or have had agricultural occupations. This kind of representation ensures that rural areas always receive their share of government development money. Rural electrification followed quickly on the heels of the opening up of the great forest areas for small dairy farms in the 1880s and 1890s, and rural roads and other infrastructures were pushed through difficult terrain to link small settlements which in many more populous countries would have been considered insignificant or uneconomic. Similarly, rural areas in New Zealand have always fared well in terms of education and health facilities. Rural hospitals, maternity hospitals, district nurse services, dental clinics, as well as an excellent tradition of rural schools, enhanced by a 'country service' system necessary for the advancement of teachers, have meant that in New Zealand rural people have suffered little in comparison with their urban counterparts. In the provision of other facilities such as the Country Library Service or the rural school swimming baths the rural vote has carried great influence in Parliament. Rural areas have never been perceived as 'problem areas', and this probably accounts for the fact that so little rural sociology has been written in New Zealand. Agriculture never had to evolve through a feudal or peasant stage; rather this was a transplant economy where farmers figured quite prominently in most indices of socio-economic status (Vellekoop 1969).

In the early 1970s a combination of adverse climate and more difficult market conditions for New Zealand's agriculture forced the raising of consciousness of rural social issues by drawing attention to rural problems. Britain's entry into the European Economic Community heralded the departure of what became known as the 'green years' for New Zealand farming (Salinger 1979), the Agricultural Production Council commissioned a report on rural social conditions in New Zealand (Lloyd 1974), and analysis of the 1971 census data revealed that rural depopulation was quite a major problem, with 72 out of the 109 counties suffering falling population in the 1966–71 intercensal period. The evaluation of these data unleashed considerable academic and political concern about the consequences of the rural exodus. For example, it was argued that the retention of people on farms and in rural areas was 'simple justice' because without a minimum level of population they could not have the welfare and other facilities that New Zealanders elsewhere had by right. Also, it was argued that people were an important factor influencing agricultural production, because smaller units produced more per hectare, and that total production in New Zealand was vital.

> It is precisely because we have overstressed the importance of economic and technical forces and underplayed the social forces at work in agriculture that we found ourselves in a situation where we need greater income yet are faced with falling output. (Morton, 1975: 85)

Note that the reason for these initiatives was that it was feared that rural social conditions might have an adverse effect on farm production, not that there was any intrinsic interest in the plight of rural people.

During the 1970s farm profitability declined in New Zealand; consequently the volume of production remained static, with farmers unwilling to invest and making efforts to cut costs to conserve net income. This drop in production caused the government to investigate ways of increasing productivity – principally subsidies, income support and changes to the taxation system. At some points official concerns coincided with the frustrated ambitions of aspiring farm owners, expressed politically through young farmer and farm worker organizations. There was increasing concern that farms were enlarging and amalgamating rapidly, leading to a loss of farm owners and farm workers (Smit 1975); that, because there was limited opportunity for new farmers, the farm population was ageing and that old farmers were less productive (Willis 1980); and that limited entry into farming meant the closing of the 'agricultural ladder' in New Zealand, apparently leading to more farms being inherited and passed on within families rather than being available for 'new blood' (Watson & Cant 1972, Willis 1982).

In general terms it is possible to agree with Cloke (1986) that during the 1970s there were some signs of a move away from agriculture and production-dominated rural studies to research oriented towards problems and people. Since the now famous 'Why they left Eketahuna' study, the chronicle of the decline of a small rural town which received much attention only because it was the first in the field (Glendinning 1978), rural population studies have been set on a firmer and much more systematic foundation (see, e.g., Heenan 1979, O'Neill 1979a, Neville 1980, Cloke 1983).

The wider publicity given to rural depopulation and the systematics of rural demography led logically to an interest in other rural social problems. A series of rural seminars was held in the late 1970s, prompted particularly by the threatened closure of maternity hospitals in places like Hunterville and Greytown. A number of more general rural social surveys focused on issues like rural housing, education and services (Sparrow 1979), the rôle of women in rural areas (Canterbury University Sociology Department 1976), the clergy in rural areas (O'Neill 1979b) and alternative employment for rural people (Gillies 1980).

It is probable that this growing interest in rural social problems, rural people and the non-agricultural elements of rural society would have continued to grow more rapidly had it not been for yet another major change in New Zealand's political economy.

The presentation of the 1984 Budget marked something of a watershed in the history of New Zealand's farming and land-use policy in that the newly elected Labour government signalled its clear intention to make the marketplace the main regulator of land-use and farm production decisions. The main reason for this change appeared to be the determination of the government and its treasury advisers to adopt a radically different economic philosophy to deal with New Zealand's generally unfavourable macro-economic situation, especially the unprecedented rates of government spending, much of it on farming and overseas borrowing to finance that expenditure. Principally, the change in policy involved the removal of farm income supports and subsidies which had long been part of the New Zealand farming scene but which had grown alarmingly in the late 1970s and early 1980s:

In our view the approach which seeks to ensure 'adequate' farm incomes is one that has inhibited the adjustment process by protecting farmers from the realities of markets. This has had a major cost to the nation and indeed has been to the longer term disadvantage of the farming community (The Treasury 1984: 1)

During the 1970s, when it became clear that total agricultural production was static, a number of rural policies to boost stock numbers and production were put in place. A Livestock Incentive Scheme to promote greater livestock numbers and a Land Development Encouragement loan scheme to stimulate the development of marginal land were the two main policies. These were followed after the 1981 election by the Supplementary Minimum Price Scheme, a subsidy policy that pegged minimum prices much further from market prices than had ever previously been the case. The general result of these policies was that land prices rose sharply whereas net farm incomes remained static or fell in real terms. At one point in the late 1970s it was calculated that the net income of the average sheep farmer was $12 900 NZ; of this, $12 100 NZ was taxpayers' money and this when the gross income was close to $100 000 NZ (Ministry of Agriculture and Fisheries Statistics 1979). With hindsight it is easy to see that subsidies, land values and incomes were getting seriously out of balance and that the crunch was just around the corner.

In fact, farmers themselves had accepted high support prices rather uneasily and many changes contained in the Labour government's new 'market forces' doctrines had begun several years before, albeit in a rather *ad hoc* fashion. One of the first manifestations of the change from traditional grassland farming was the emergence of smallholdings, especially around the fringes of major metropolitan areas. This trend helped to stem the depopulation of rural areas which had occurred so heavily in the 1960s and early 1970s (see Heenan 1979). The other major agricultural diversification which took place before the 1980s was the development of horticulture – mainly the growth of kiwifruit and other subtropical fruit, which occurred mainly in land previously occupied by dairy farms. This development brought greater intensification of settlement to some rural areas and the demand for more seasonal labour (Stokes 1983). Both the growth of peri-urban smallholdings and the expansion of small horticultural units initially created major planning problems for conservative rural, farmer-dominated local authorities. These authorities viewed with suspicion the hippies and alternative lifestylers moving into their communities, who were very hung up with 'minimum economic unit size' planning concepts (see Cant 1980).

The point being made is that the refocusing of national and rural attention on agricultural problems because of the economic difficulties of farmers was a setback to rural research on non-agricultural topics. The 'crisis' in agriculture began to appear in late 1985; the combined effect of the removal of subsidies, the appreciation of the new floating New Zealand currency and world surpluses for agricultural products sent farm incomes plummeting and land prices tumbling for the first time in several decades. Those who suffered most were farmers who had purchased land at highly inflated values and saw their equity in their property fall by up to 50% almost overnight. Naturally enough this crisis seemed to focus research attention on the plight of rural communities and the small rural towns dependent on farming – especially the popular press and pressure group research. Farmers marched on Parliament for almost the first time in New Zealand's history; then demanded a better deal from government and there were dire predictions of thousands of farmers being forced to walk off the land in 1930s depression style.

In fact, the publication of the 1986 Census of Population and Dwellings revealed that a range of people-oriented and rural planning problems had been emerging, even though they were disguised by the attention given to farming issues: the confirmation of farm enlargement and the spectacular reduction in the farm labour force; the decline of smaller rural servicing centres; the continued growth of horticulture-related activities; the expansion of tourism in rural areas; the impact of the closure of rural industries such as dairy factories and freezing works; and the rapid growth of beach retirement centres in the 'sunbelt' of the North Island. All these are planning issues affecting both the farming and non-farming sections of the rural population. I shall now discuss these issues in more detail.

The rural population associated with farming

Cloke (1986) emphasized a growing recognition in New Zealand rural studies of the distribution between the farming and non-farming elements of the rural population. The definition of rural and urban populations employed in the New Zealand Census of Population and Dwellings dates back over 100 years, when the rural population was first clearly defined as 'persons dwelling in counties or small towns with a population of less than 1000'. Table 10.1 shows the urban–rural distribution of the New Zealand population and the intercensal increase for 1926–81. The final figure for the 1986 census was not available at the time of publication.

A more detailed breakdown of the characteristics of the rural population for the 1976 census reveals that only 53% of the male labour force were employed in agriculture, forestry or hunting occupations and only 3% of the female labour force (Department of Statistics 1983: 31).

Obviously the term 'the rural population' includes a wider range of people than simply those involved in farming. When the spatial aspects of rural population change are considered it is possible to extract several diverse trends related mainly to the distinction between farming elements and non-

Table 10.1 Urban–rural[a] distribution of new New Zealand population, and intercensal increase, 1926–81[b].

Census	Urban (number)	Percentage of total population	Rural[c] (number)	Percentage of total population	Percentage change Urban	Rural
1926	952 102	67.9	449 572	32.1		
1936	1 065 228	67.9	503 885	32.1	11.9	12.1
1945	1 227 069	72.2	472 076	27.8	15.2	−6.3
1951	1 424 745	73.7	508 849	26.3	16.1	7.8
1956	1 625 887	74.9	543 727	25.1	14.1	6.9
1961	1 866 894	77.5	542 525	22.5	14.8	−0.2
1966	2 145 601	80.3	526 507	19.7	14.9	−3.0
1971	2 361 314	82.6	496 171	17.4	10.1	−5.8
1976	2 614 119	83.6	511 004	16.4	10.7	3.0
1981	2 650 904	83.6	520 487	16.4	1.4	1.9

Notes

[a] The urban population has been defined as that in main and secondary urban areas, as well as that of all other towns of 1000 population and over. The rural population is that not defined as urban.

[b] Populations at earlier censuses have been adjusted to boundaries existing at the 1981 census date.

[c] Excludes shipping.

Source: Department of Statistics (1983: 14).

farming elements. This is especially so for traditional grassland farming (dairying and sheep and beef cattle raising) as distinguished from horticulture and small farming activities. Figures 10.1 and 10.2 show rural population increase and decrease for the 1981–6 intercensal period. In addition to counties, districts and rural centres with less than 1000 population, small towns of 1000–4999 are included where they have *increased* more than 25.2% (six times the national average) or decreased more than 10%. Figure 10.2 shows that, with one or two minor exceptions, all counties lost population, except where affected by tourism, rural–urban periphery subdivision, horticultural subdivision or beach resort subdivision.

Traditional dairy farming counties and traditional sheep and cattle farming counties have continued to lose population due to a number of interrelated processes: farm amalgamation or enlargement, reduction of farm labour as a cost-saving measure, and the closure of rural processing industries. Table 10.2 shows changes in the number of holdings by farm type for the 10 years from 1973 to 1983. It is clear that some types of holdings have suffered marked declines in numbers, especially dairy farmers (−135%). Although the number of traditional livestock farms grew by 6% over the decade, they actually decreased as a percentage of the total num-

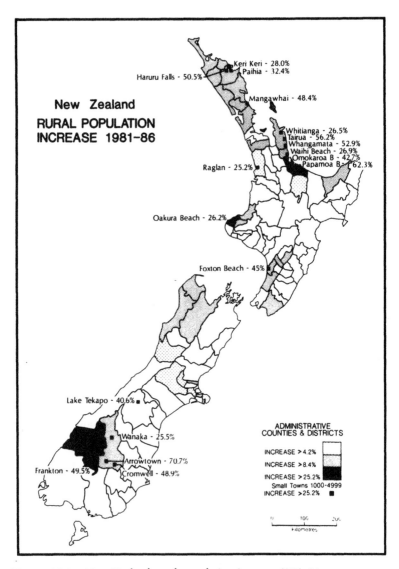

Figure 10.1 New Zealand rural population increase 1981–86.
Source: New Zealand Census of Population and Dwellings 1986.

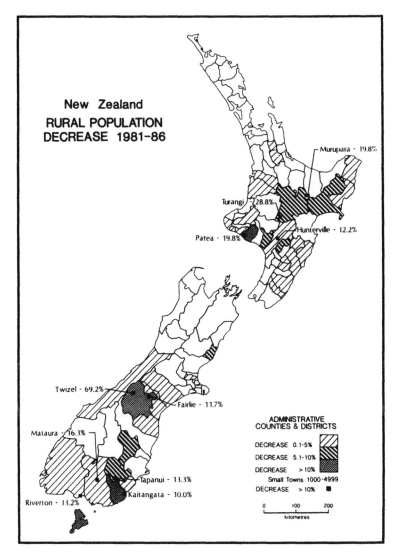

Figure 10.2 New Zealand rural population decrease 1981–86.
Source: New Zealand Census of Population and Dwellings 1986.

Table 10.2 Changes in number of holdings, 1973–83.

Farm type	Change 1973–83		Number 1973	Percentage of holdings	Number 1983	Percentage of holdings
all dairying	−2443	(−135%)	18 144	28.7	15 701	20.7
all sheep	+6307	(+28.9%)	21 822	34.5	28 129	37.1
all beef	+539	(+7.1%)	7602	12.0	8141	10.7
all pig	−30	(−4.6%)	655	1.0	625	0.8
all cropping	+565	(+34.6%)	1631	2.6	2196	2.9
poultry	−191	(−30.1%)	635	1.0	444	0.6
other mixed livestock	−1211	(−23.2%)	5215	8.3	4004	5.3
total traditional	+3566	(+6.0%)	55 674	88.1	59 240	78.2
market gardening	+114	(+7.0%)	1623	2.6	1737	2.3
orchards	+466	(+22.7%)	2054	3.3	2520	3.3
plantations	+380	(+92.5%)	411	0.7	791	1.0
other[a]	+8016	(+235.3%)	3406	5.4	11 422	15.1
total	+12 549	(+19.9%)	63 196	100.0	75 745	100.0

Notes

[a] includes mushroom growing; grape growing; berry fruit; tobacco; hops; flowers; plant nurseries; beekeeping; orchids; other fruit and vegetables; other farming; idle land.

Source: Agricultural Statistics, 1973–83.

ber of farms from 88% to 78%. Conversely, the big growth in the numbers of farm holdings came in horticulture and smaller intercensal land uses (235% increase).

These changes in the number of farms are reflected in changes in the number of people involved in farming. The Valuation Department records that between 1973 and 1983 over 4500 dairy farms were sold in New Zealand for the purposes of farm enlargement, i.e. where previously those farms had supported a family now they do not because they form part of another farm. Table 10.3 shows the distribution of this loss of labour from rural areas; for example it shows a net loss of 4239 male working owners and 10 646 male permanent employees in the ten year period 1973–83. This is again contrasted with a net gain of 1277 working owners in the 'other' category and a net gain of 3168 permanent employees. Note that it is unwise to use the same data for females because tax changes during the period dramatically increased the number of women classifying themselves as owners.

The reduction in the number of farmers and full-time agricultural workers has in some areas been exacerbated by the closure of rural process-

Table 10.3 Changes in employment on farm holdings, 1973–83 (males only).

Principal farm type	Working owners				Permanent employees			
	1973	1983	Change Number	%	1983	1983	Change Number	%
dairy	19 141	16 810	−2331	−12.2	6624	2696		−59.3
							−3928	
sheep	18 653	20 833	+2230	+12.0	10 983	7720	−3263	−29.7
beef	5588	4011	−1577	−28.2	1665	849	−816	−49.0
pig	558	446	−112	−20.1	314	268	−46	−14.6
cropping	1112	1522	+410	+36.9	354	398	+44	+12.4
poultry	543	377	−166	+30.6	453	277	−176	−38.9
mixed livestock	4888	2195	−2693	−55.1	3349	888	−2461	−73.5
total traditional	50 483	46 244	−4239	−8.4	23 742	13 096	−10 646	−44.8
market G	1391	1342	−49	−3.5	562	551	−11	−2.0
orchards	1751	1824	+73	+4.2	857	1025	+168	+19.6
other	2429	3706	+1277	+52.6	2921	6089	+3168	+108.5
total	56 133	53 116	+3017	+5.4	28 082	20 761	−7312	−26.1

Note
Agricultural Statistics, 1973–83.

ing industries, particularly the amalgamation and concentration of dairy factories and in some cases the closure of meat-freezing works. The dairy farming region of Taranaki on the west coast of the North Island of New Zealand provides a classic case of the concentration of dairy factories. Table 10.4 shows some of the trends nationally in the dairy industry in the 10 years 1973–83, and these include the loss of 53 Co-operative Dairy Companies, the loss of 67 dairy factories, a loss of 3781 factory suppliers, but an increase in average herd size from 106 to 137 cows.

Figure 10.3 shows geographically the spatial concentration of dairy factories in Taranaki; what needs to be remembered is that in most cases the dairy factory represented the centre of a rural community which often consisted of a general store, factory workers' houses and sometimes a garage. The closure of so many factories has also led to a major reduction in unskilled employment opportunities for Maori people, who comprised a large percentage of the traditional dairy factory workers, as they did in meat-freezing works.

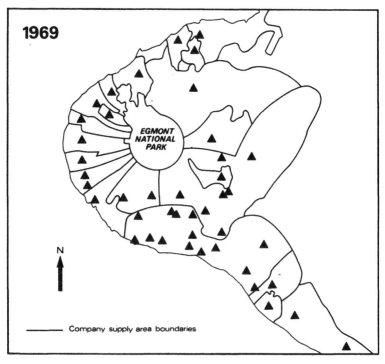

Figure 10.3 The location of dairy factories in Taranaki 1969, 1975, 1983.
Source: New Zealand Dairy Board.

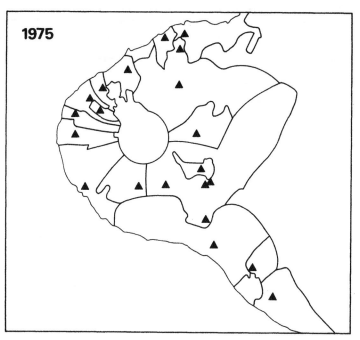

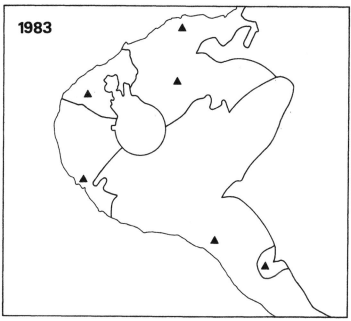

Table 10.4   Trends in the New Zealand dairy industry, 1973–83.

Item	1972–3	1982–3	Actual change	Percentage change
factory suppliers	18 226	14 445	−3781	−20.0
average farm size (ha)	106	137	+37	+29.3
number of co-op companies	89	36	−53	−59.6
butter factories	53	28	−25	−47.2
cheese factories	75	27	−48	−64.0
milk powder factories	26	32	+6	+23.0
average output in cheese	2655	6023	+3368	+126.9
kilometres travelled by milk tanker (millions)	26.5	28.0	+1.5	+5.7

Source: Agricultural Statistics, 1973–1983.

The combined impact of these changes to the farming-related rural population is clearly illustrated with a small community case study, again from Taranaki. Figure 10.4 shows that, in 1960, 13 dairy farms produced for a small dairy factory serviced by a local general store. By 1980 the factory was closed, the store was closed and the factory workers' houses removed; through farm amalgamation the 13 farms had been reduced to 10. Similarly, in 1960, 85 children caught the school bus to the local nearby town, but in 1980 only 14 children attended school. The demographic change was, of course, produced by a number of factors, including smaller family size and fewer farms.

Policies and plans dealing with the problems affecting the farming sector of the rural population have varied according to which party is in power in New Zealand. For example, when the problems of rural depopulation and farm amalgamation were publicized in the early 1970s the Labour party pledged itself to lend less money for farm enlargement through the Rural Bank. When Labour achieved power in 1972 they did carry this policy through and began to encourage new farmer settlement. However, their 'more market' economic philosophy outlined earlier, which the current Labour government has put in place, has cut cheap lending for all purposes, so it is apparent that farms will continue to get bigger. This policy making has been dominated by the overall economic policies outlined earlier, and there have been precious few policy or planning initiatives dealing with the social problems facing farming people. Part of the problem in New Zealand is the small numbers involved. Take, for example, the problem of employment retraining for displaced farm workers. In 1986 the Labour government abolished the Agricultural Training Council as part of its cost-cutting measures but charged the Vocational Training Council with the

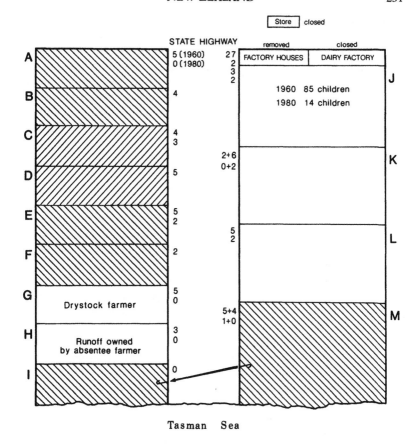

Figure 10.4 Changes in rural Taranaki 1960–80.
Source: Author's research.

task of devising retraining programmes for farm workers displaced during the rural crisis. However, when the Vocational Training Council tried to discover who needed training in the King Country region – a region of hard hill country farms – it was found that only about 2% of the 600 farms actually employed full-time labour (Victoria University Geography Department 1986). So the number of people liable to be affected by this policy was very small – it is difficult to bring down a policy for 20 people.

The non-farming rural population

The general trends in New Zealand's rural population shown in Figures 10.2 and 10.3 need to be explored in more detail to isolate the trends affecting the non-farming sector of the rural population. By concentrating on small towns (1000–4999) it is possible to highlight several major themes. Table 10.5 divides up these smaller towns by greatest increase and decrease in population (1981–6). It further classifies them according to type:

Group 1 are the beach/retirement towns. Most of them are on the East Coast in the northern part of North Island. As the population of New Zealand ages and as people increasingly desire the warm climates and beach lifestyles these centres will continue to grow.

Group 2 are those towns which have grown steadily with the expansion of the kiwifruit industry, where the subdivision of properties, the demand for seasonal labour, and the general injection of capital has pushed expansion.

Group 3 are towns benefiting from tourism in South Island, with settlements like Arrowtown and Frankton near Queenstown recording quite dramatic population increases, the influence of both summer and winter attractions.

Group 4 are the traditional small rural servicing centres which have declined with the depopulation of the farming areas as services have centralized and communication improved. Only 5 of the 14 towns are in North Island; the other small towns are all south of Christchurch in the southern part of the South Island. This reflects the general northward drift of the New Zealand population.

Group 5 are towns where a particular industry or infrastructure has influenced the population decline. The classic examples are the completion of construction of hydroelectric power schemes affecting Turangi (central North Island) and Twizel (central South Island) or the closure of the Patea freezing works (West Coast North Island).

Another influence on the non-farming element of the rural population which does not appear in Table 10.5 is the so-called Think Big major projects industrialization. In the late 1970s the National party government decided to use the nation's energy resources to implement a major industrialization programme based on becoming more self-sufficient in energy. These projects included major expansions to the country's only oil refinery, expansions to the country's only steel mill, the construction of a $2 billion synthetic petrol plant using natural gas, an export methanol plant, and an ammonia urea fertilizer plant. In addition, it was decided to electrify the main trunk railway in North Island and build several small

Table 10.5 Population trends in New Zealand towns of 1000–4999 in 1981–6 intercensal period.

Towns increasing 25.2% or more		Towns decreasing 5% or more	
1 Beach/retirement towns		**4 Rural servicing centres**	
Raglan	25.2	Taihape	5.5
Oakura	26.2	Darfield	7.0
Whitianga	52.9	Balclutha	6.5
Tairua	56.2	Paeroa	6.9
Omokaroa Beach	42.7	Edgcumbe	5.9
Papamoa Beach	62.3	Hunterville	12.2
Foxton Beach	45.0	Weston	5.1
Paihia	32.4	Wyndham	8.6
Haruru Falls	50.5	Otautau	7.6
Waihi Beach	26.9	Tapanui	13.3
Mangawhai	48.4	Lawrence	8.7
		Piopio	8.1
2 Kiwifruit towns		Fairlie	11.7
Te Puke	25.8	Roxburgh	5.8
Kerikeri	28.0		
Katikati	27.0	**5 Rural towns with declining industry**	
		Murupara	19.8 (forestry)
3 South Island tourist towns		Turangi	28.8 (hydroelectric
Arrowtown	70.7		scheme)
Lake Tekapo	40.6	Twizel	69.2 (hydroelectric
Frankton	49.5		scheme)
Wanaka	25.5	Mataura	16.3 (flood damage)
Cromwell	48.9	Moerewa	5.9 (freezing works)
(also hydro electric scheme)		Patea	19.8 (closed freezing
			works)
		Kaitangata	10.0 (coal mining)

Source: NZ Census of Population and Dwellings, 1986.

forest-based processing plants. The point about these major projects is that, especially during the construction phase, several of the projects had big impacts on rural areas. For example, the three Taranaki gas-based projects were set in rural areas close to the major regional centres of New Plymouth, leading to a tremendous population boom for New Plymouth and the surrounding counties of Taranaki, Clifton and Inglewood. These projects had the added significance of prompting central and local government to become deeply involved in social, economic and environmental impact studies in a way never before experienced in New Zealand, spawning a whole new literature on the subject and forcing regional planners to grapple with the problems of major industries in rural areas (see Speers 1978, Taranaki United Council 1982–3, Taylor *et al.* 1983). The National party government had brought the whole planning process into sharp political focus by passing the National Development Act, a 'fast-track' procedure which attempted to sidestep some provisions of the existing planning legislation. The promotion of the Bill and its passage through Parliament drew storms of protest and led to the formation of a political watchdog group, the Coalition for Open Government (COG), whose research provided the much needed fuel for a healthy debate on the planning and impact assessment of these projects. For example, the threatened spoiling of Maori fishing grounds by the outfall from one of the Taranaki petrochemical plants led to an appeal by local Maoris for the provision of the Treaty of Waitangi (1840), guaranteeing their fishing resources. This set a successful precedent which has since been used several times in rural areas.

Again the point should be made that it is primarily national economic and social policies that are affecting rural areas, rather than policies designed specifically for rural problems. Another example is the recent stress by the Labour government on restructuring government departments into public corporations, with emphasis on efficiency and public accountability. For some government departments this has had a profound effect on rural areas. The corporatization of New Zealand Railways, for example, has meant the loss of several thousand jobs nationally, with significant numbers in smaller rural centres like Taumaranui and Te Kuiti. The centralization of offices of the New Zealand Forest Service and the Department of Lands and Survey is currently having a similar effect, the dimensions of which are not yet fully recognizable. Thus the government's macro objectives of reducing public expenditure and increasing public sector efficiency conflict with the idea of keeping rural areas viable and make it difficult for the same administration to put in place policies which may ameliorate the situation. As yet there are few specifically rural policies to cope with government sector restructuring, apart from policies directed toward farming.

One set of policies which in New Zealand's case are important for rural areas are regional development policies. Regional economic differences are traditionally very small in New Zealand (Williamson 1968). Regional Development Priority Areas have existed since 1973 and consist mainly of the more isolated non-metropolitan regions where the economic base is strongly rural. Table 10.6 shows the dimensions of the Regional Assistance Programme from its inception in 1973 until 1986, when there was a major policy change. The programme in New Zealand basically follows a policy direction recently stated by the OECD in assisting mainly small business:

The future development of rural areas may well depend on the viability of small and medium sized enterprises which despite their high failure rate still create large numbers of jobs and demonstrate considerable initiative, innovativeness and flexibility in running their businesses. (OECD, 1986: 10)

In 1986, again as part of government expenditure cuts, the regional assistance programme was changed in two ways. Firstly, the various loan schemes have been suspended and replaced by investigation grants. Secondly, these investigation grants now apply to all regions, demonstrating that where previously regional development policy may have given some advantage to rural areas, now it does not. Indeed, recent research shows that regional development loan assistance was quite a good investment, especially compared with the cost of job creation in the Think Big programme or the cost of paying unemployment benefit (Willis 1986: 16). Table 10.7 compares the cost of job creation in the rural King Country region with the cost of jobs in the Think Big projects.

Table 10.6 Regional development assistance, 1973–86 (March years).

Region	Total assistance NZ $	% total	Population 1986 (000s)	% Population 1986
Northland	4 896 053	9.7	127 864	14.6
King Country	1 616 165	3.2	32 230	3.7
East Coast	3 223 906	6.4	54 070	6.2
Taranaki	4 656 813	9.2	107 503	12.3
Wanganui	4 621 730	9.1	70 513	8.0
Wairarapa	1 170 617	2.3	39 601	4.5
total North Island	20 185 284	39.9	431 781	49.2
Marlborough	2 886 182[a]	5.7	38 239	4.4
Aorangi S. Canterbury)	5 868 868[a]	11.6	81 760	9.3
West Coast	5 715 495	11.3	34 961	4.0
Otago	8 468 078	16.8	186 039	21.1
Southland	5 166 181	10.2	104 458	11.9
total South Island	28 104 804	55.6	455 482	50.8
		100.0		100.0
total New Zealand	49 290 088 + 2 264 836 = 50 554 924		877 263	

Notes
[a] Not full period.

Total other NZ $2 264 836. Region not priority.

Sources: Department of Trade and Industry files; NZ Census of Population and Dwellings, 1986.

Table 10.7 Comparison of the cost per job in Think Big projects and King
Country manufacturing projects.

Project	Total cost per job NZ $	37% government contribution assumed NZ $
Marsden refinery	9 230 000	3 415 110
Clyde Dam	17 400 000	6 438 000
NZ Steel, stage II	575 806	213 049
main trunk electricity[a]	24 661 000	9 124 570
Waikato coal development	244 012	90 213
Martha Hill mine	966 101	357 457
Ammonia Urea production	986 000	2 666 666
King Country manufacturing		5474

Notes
[a] Figure takes no account of the net job loss.

1   Jobs in the construction phase not included.
2   37% rate of government contribution to major projects has been assumed in order to
    estimate total cost of jobs. In fact, the government may have to pay much more, e.g.
    Ammonia Urea.

Source: Major Projects Advisory Council newsletters.

The discussion so far has emphasized the domination of central government in affecting rural plans and policies in New Zealand, barely mentioning the rôle of regional institutions or local authorities. That this emphasis
is substantially correct was crudely demonstrated in November 1986 when,
under provisions in the mining and prospecting law, a government prospecting team broke down fences and began drilling for coal without the
private landholder's permission, causing a storm of protest among the
rural community.

New Zealand has, in fact, evolved a full list of united and regional councils
in the last five years, but they have very little power to collect taxes (rates)
and the planning schemes they may bring into place are largely indicative
rather than compulsory. Only the Auckland Regional Authority is elected
directly by all members of the region, and the major function of these
bodies is co-ordination of the functions which are wider than their constituent local authorities, especially urban transport and civil defence.

Local authorities such as counties and districts have more statutory planning power to affect rural policies, but the land tax they collect allows them
to administer a limited range of functions – things such as land use, water
supply, waste disposal and roading. Any people-directed aid invariably

comes through the local branches of state departments, such as Social Welfare, the Health Department, the Department of Maori Affairs and the Labour Department.

Such is the influence of central government policies that, by the end of 1986, a noticeable dualism was emerging in national and regional economic health in New Zealand. The free market policies of the Labour government had led to an explosion of economic activity in the big cities, fuelled by property investment and record stock exchange activity, with the lowest unemployment levels in Auckland and Wellington. Meanwhile, unemployment had grown to near record levels in the provincial centres and rural areas as the impact of central government's deregulatory policies hit the farming industry and associated rural services.

Conclusions

It has been a major theme in this chapter that, although some signs of rural social research and rural policies which were people- and problem-oriented had begun to emerge in the late 1970s, the recent dramatic changes in government macro-economic policy, affecting both farming sector and government restructuring, have tended to obscure both rural social research and rural social policy. The so-called farming crisis and its effect on small rural servicing centres have directed attention away from other emerging groups in the rural population, namely part-time farmers and retirement/beach settlements, tourist settlements and rural industries, postponing the development of policies specifically directed towards rural people and rural areas, with even regional development policy which previously favoured rural areas being widened to include metropolitan areas. These developments have been largely ushered in by a Labour government with a predominantly urban support base. The test of the political sustainability of these policies will come in the 1987 General Election, when it is certain that the National party opposition with its strong rural support base will foreshadow a wide range of rural policies, including some people-oriented, problem-oriented rural social policy.

References

Cant, R. G. 1980. *People and planning in rural communities. Studies in Rural Communities No 4.* Canterbury University, Christchurch.

Canterbury University Sociology Department 1976. *The rural women of New Zealand: a national survey.* Canterbury University, Christchurch.

Cloke, P. J. 1983. Policy responses to rural depopulation and repopulation contracts between New Zealand and the United Kingdom. *New Zealand Agricultural Science* **17** (1), 231–6.

Cloke, P. J. 1986. Observations on policies for rural communities in New Zealand. *New Zealand Geographer* **42** (1), 2–10.

Department of Statistics 1983. *Rural profile*. Occasional Paper No. 4. Wellington.

Gillies, A. 1980. Alternative employment opportunities in rural areas in New Zealand. *New Zealand Agricultural Science* **14**, 9–17.

Glendinning, D. 1978. *Why did they leave Eketahuna?* Wairarapa Education and Rural Services Committee. Eketahuna.

Heenan, L. O. B. 1979. The demographic condition of New Zealand rural society. In *New Zealand rural society in the 1970's*, R. Bedford (ed.), 2–22. Studies in Rural Change No. 1. Canterbury University, Christchurch.

Lloyd, D. W. 1974. *A preliminary review of rural social conditions with particular reference to the manpower position on farms*. Ministry of Agriculture and Fisheries, New Zealand Agricultural Statistics 1979. Wellington: Agricultural Production Council.

Morton, H. A. 1975. People as a factor influencing future agricultural production. *New Zealand Agricultural Science* **9** (3), 85–90.

Neville, R. J. 1980. Spatial patterns of population change: trends in rural areas and small towns. In *The land our future: essays on land use and conservation in New Zealand*, G. Anderson (ed.), 261–90. Auckland: Longman.

OECD (Organization for Economic Cooperation and Development) 1986. Managing the rural renaissance. *OECD Observer* **141**, July, 9–12.

O'Neill, C. J. 1979a. New Zealand's population: trends and implications for the rural community. *New Zealand Agricultural Science* **13** (1), 10–17.

O'Neill, C. J. (ed.) 1979b. *Interstices: a report on the consultation on rural ministry Waimate*. Studies in Rural Change No. 3. Canterbury University, Christchurch.

Salinger, J. 1979. *Climatic change: lessons from New Zealand's historical records*. Paper presented to the 49th ANZAS Conference, Auckland, NZ.

Smit, B. A. 1975. An analysis of the determinants of farm enlargement in Northland, New Zealand. *New Zealand Geographer* **31** (2), 160–77.

Sparrow, M. *et al.* 1979. *Banks Peninsular 1977: a rural survey*. Studies in Rural Change No. 2, Canterbury University, Christchurch.

Speers, R. S. 1978. *Influence of industry in a rural setting: New Zealand steel and Waiuku Borough*. MA. Thesis in Geography, Auckland University.

Stokes, E. 1983. *The impact of horticultural expansion in the Tauranga District*. Town and Country Planning Technical Report No. 14, Ministry of Works, Wellington.

Taranaki United Council (1982–3) *Energy monitor: study of the impact of petrochemical projects on the work force and accommodation in the Taranaki region*. Reports 6–8, New Plymouth.

Taylor, C. N., C. Bettesworth & J. G. Kerslake 1983. *Social implications of rural industrialisation: a bibliography of New Zealand experiences*. Lincoln College Centre for Resource Management, Christchurch.

The Treasury 1984. *Economic management land use issues*. Government Printer, Wellington.

Vellekoop, C. 1969. Social strata in New Zealand. In *Social process in New Zealand*, J. Forster (ed.), 233–71. Auckland: Longman.

Victoria University Geography Department 1986. *Survey of farmers in Taumarunui County*. Unpublished.

Watson, M. B. & G. Cant 1972. Variations in productivity on Waikato dairy farms. In *Proceedings of the New Zealand Geography Conference*, 165–75. Christchurch.

Williamson, J. G. 1968. Regional inequality and the process of national development: a description of patterns. In *Regional Analysis*, L. Needleman (ed.). Middlesex, England: Penguin.

Willis, R. P. 1980. Dairy farmers younger yet. *New Zealand Journal of Agriculture* July, 22–4.

Willis, R. P. 1982. The Influence of some social factors on agricultural production, *New Zealand Journal of Agricultural Science*, **16** (3), 160–4.

Willis, R. P. 1986. Manufacturing in the King Country. In *King Country Regional Resources Survey* (forthcoming), Victoria University of Wellington.

11   Conclusions: Rural policies – responses to problems or problematic responses?

PAUL CLOKE

Scale and diversity

The preceding case studies were selected so as to provide a range of different geographical and sociopolitical contexts in which to examine policies and plans for rural people in the developed world. These contexts range from small and compact nation–states to vast subcontinents with high levels of internal diversity. The accounts of rural policies reflect these differences of scale, and although terms such as 'peripheral' and 'marginal' have been coined in each of the countries under scrutiny, there are significant variations in marginality to be acknowledged, for example between the north of the Netherlands and northern Canada or peripheral areas of the USSR. Remoteness should, therefore, be recognized as at least partly a scale-determined phenomenon, and the consequent range of scales on display in this book is liable in many ways to skew comparative analysis of remote rural areas.

Similarly, the information presented in the case studies involves variation in sociocultural mix, and this factor should again be accounted for in validating any comparisons and contrasts arising from evaluations of rural policies in different nations. Thus the marginal Welsh and Scottish cultures within the British context might more readily be compared with French-speaking populations in Canada than with the 'marginal' Inuit nations in the Canadian north, or the Siberian peoples in the USSR.

Obvious political differences also arise from these studies. Leaving aside the distinctive political economy of the USSR, which contrasts starkly with other systems of government discussed in this book, there are a range of political parties and structures presented to us here. Conservative, Republican, National or Labour parties may be caricatured but will not be easily characterized in an international context. For example, the 1984 Labour party government in New Zealand appears to exhibit many of the free-market tendencies of Conservative governments in Britain and those under other titles in the Netherlands, the USA and elsewhere. Political diversity also occurs in the tiers of government which have been established in different nations. There are clear policy-making implications arising from a federation of states or provinces where 'central state' functions are divided, as compared with the central/local government divisions found in (often smaller) nations.

This diversity between the contexts presented in the case-study chapters in this book might be used as a legitimate excuse for not indulging in any analysis of comparison or contrast between them. Indeed the difficulties of such an analysis should not be underestimated. Nevertheless, if we are to break down the parochialism which currently besets our understanding of planning and policy making, indications of communality and comparability in an international context are a necessary step, albeit perhaps a tenuous one, towards more widely applicable explanations of rural change and the responses of governments to such change.

A few preliminary conclusions are presented in this final chapter. These are of necessity both brief and tenuous, yet it is hoped that this book will act as some small catalyst to a greater urgency for international comparative study of planning and policy making in rural areas. As explained in Chapter 1, authors of case study chapters in this book were presented with a common agenda relating to the problems experienced by rural people, power relationships associated with planning for rural people, the planning mechanisms which have been established, and the relationships between implementing agencies. It was hoped in this way that case-study evidence would illustrate:

(a)   any communality of problems experienced by particular fractions of the rural population;
(b)   any comparability in the planning and policy mechanisms which have been developed in response to those problems;
(c)   any supportive evidence of contemporary concepts of the rôle of planning as a state activity.

These issues are now dealt with in turn.

Communality of problems

The characteristic problems of rural areas described by the various authors in this book reflect the diversity of scale, culture and political economy of the nations under scrutiny. Yet a very strong underlying theme emerges from the analyses presented here. Rural areas are in most cases viewed as synonymous with old resource regions and are thereby imbued with symptoms of economic underdevelopment. Thus Gerald Hodge describes non-urban regions in Canada as displaying poverty, illiteracy, poor housing and public infrastructure, obsolete resources and inefficient technology. Such problems are noticeable at both regional and local scales. In the larger nations, where whole regions are marginal in political and economic as well as geographical terms, these indications of relative deprivation apply wholesale. Elsewhere, even amongst densely populated and relatively accessible rural areas such as those in Britain, the Netherlands and parts of France, the problem of deprivation persists, but is interspersed with the wealth of more adventitious residents of the same countryside. In Britain, for example, there appear to be specific groups of rural people experiencing deprivation, including the remnants of the former agricultural economy of the area; the *nouveaux pauvres* who have consciously decided to live their life in rural areas for reasons of environmental choice, yet who have little opportunity of access to wealth or income to make such a life an economically prosperous one; and the elderly components of all resident fractions including the formerly adventitious.

There do, therefore, seem to be a series of endemic problems associated with the decline of extensive resource-based economies in rural areas. Different case studies reflect different forms of economic restructuring in rural areas, and it is clear from the evidence presented that rural problems are not simply those connected with the capital substitution for labour within prosperous agriculture or the loss of gainful employment because of agricultural decline. There are indeed indications of policy responses which assume that rural problems stem from agricultural restructuring. In New Zealand, for example, rural policies do seem to have been synonymous with farming policies as successive governments have recognized the necessity to prop up that country's most important export industry. Similarly, Judith Pallot's account of policy in the USSR suggests that attention to rural problems has stemmed from the problematic nature of a decline in agricultural production due to the loss of rural workers to urban areas.

Nevertheless, other resource-based developments have occurred in rural areas, so restructuring the economy alongside agricultural changes and creating further rural resource-based problems. Bill Lassey, Mark Lapping and John Carlson describe such problems in the USA context as part of a boom–bust phenomenon. They refer to cycles of economic restructuring in rural communities with a heavy dependence on a narrow employment base, typically consisting of only one or two industries. Vulnerability to international variations in commodity demand means that such communities *boom* one year and *bust* the next, with tragic welfare consequences for local residents unless specific policy interventions are forthcoming. In the USA, the boom–bust phenomenon has been particularly acute in economies dependent on energy production (especially oil and coal). The recomposition of local society which follows the boom–bust restructuring has been mirrored in rural areas either side of the North Sea. One of the principal reasons behind counterurbanization trends in rural areas of north-east Scotland in the 1970s was the influx of development related to the North Sea Oil production. Now, with the slump in oil prices worldwide, communities which grew both in size and prosperity have now slumped. The description of American 'ghost-towns' – jobs lost, businesses closed, services diminished, real estate vacant and declining in value, prosperous citizens suddenly jobless or even bankrupt – refers equally to the Aberdeen area and to a lesser extent to similar communities in Norway.

These rounds of capital restructuring form a basic plank of any political economic analysis of rural areas. The development of capitalism has created economic and social structures which manifest themselves in spatial distributions, including those which are reflected in what are perceived as rural–urban differences. Spatial distributions of economic and social structures change over time as areas of advantage and disadvantage emerge and re-emerge. Gordon Cherry has illustrated this progressive substitution of dominant social and economic orders:

> Over a number of centuries the feudal order was replaced by an agrarian capitalist order. Over the last 100 years we have seen how the enfeeblement of agrarian capitalism produced the twentieth-century problems to which planning in its widest sense has reacted: land, poverty, insecurity of jobs, housing squalor, community disadvantage and restriction of opportunity. In the last 30 years new forms of conflict have arisen, and others have sharpened as one dominant value system has challenged another: the urbanite earmarks rural land for recreation, urban water needs take over rural farm land for reservoirs, the city dweller moves into rural housing. (Cherry 1976: 265).

Many of the rural areas analysed in this book reflect this substitution to one degree or another. Alongside traditional agricultural production and other primary industrial enterprises (including the more recent energy developments) there are also in some areas clear trends towards counterurbanization, including an urban-to-rural shift in manufacturing industry. In some cases, such as the United Kingdom, counterurbanization has been widespread, enfolding the most marginal areas. In Canada, the USA and Australia, there remain depressed rural areas almost untouched by this resurgence. Where rural revival is closely linked with modern manufacturing industry the vulnerability to boom–bust remains. Evidence from the USA, for example, suggests that recent downturns in specific high-tech industries have had detrimental impacts on some rural communities where problems were thought to have been solved due to an advantaged status in attracting a modern industrial base.

In areas where capital accumulation is the driving force behind change, then, people in rural localities suffer from the impacts of successive rounds of capital restructuring which tend to allocate a narrow economic base to these areas of small and scattered workforces. Restructuring can benefit rural areas, as currently in parts of Britain where restructuring has led to an influx of manufacturing industry and employment. Where such changes are not occurring through the market mechanism, governments are faced with the issue of whether to promote economic growth in rural localities. If (for whatever reason) such promotion occurs, one of the major selling points of rural areas to industrialists is the availability of a low-cost and non-unionized workforce. New employment attracted in this way may well add to the persistent problems of low income in rural areas.

Besides these issues of economic development, the other main recognizable problem in rural areas is that of providing an adequate level of services to a small dispersed population. Most chapters have discussed this issue, and the standard and cost of service delivery in rural areas is evidently a consistent issue facing rural planners and policy-makers. This theme can be approached in terms of justice and equality, suggesting that egalitarian political virtues are being brought to bear on the problem to ensure that rural people receive approximately the same servicing conditions as their urban counterparts. Alternatively, the provision of rural services can be viewed as the price to be paid for maintaining a strategic or economic presence in otherwise underpopulated areas.

Despite this clear complementarity of perceived problems in most of the nations studied in the case study chapters, albeit occurring in different configurations because of varying degrees of capital restructuring, it should be reiterated that it is the response to these problems which remains the priority of this book. The fact that similar problems occur at different scales and in different configurations suggests that if planning performs the rational, apolitical rôle discussed in Chapter 1 it will represent an attempt by government both to regulate change and specifically to intervene in problematic issues so as to respond to the needs of rural people. According to other theoretical concepts of planning and policy making, the state will be attempting to secure the needs of specific capital interests through planning, and intervention in areas of social consumption will be largely for legitimation purposes. It is to these issues of problem response that we now turn.

Comparability of planning and policy response

The various preceding case study chapters have revealed policies emanating from differing levels of government which have had significant impacts on rural people. Although not mutually exclusive, three categories may be used to convey these different policy derivations.

(1) TOP-DOWN RESPONSES

In most cases top-down policies are characterized and conditioned by the views of central government decision-makers as to how rural areas fit into the overall scheme of things. More specifically, rural areas in this context represent one (often relatively insignificant) part of the overall objectives of the state which tend to be urban in orientation and priority. Gerald Hodge's analysis of rural policy in Canada, for example, stresses that rural areas are viewed as resource environments for the dominant urban industrial society or as residual 'green spaces' awaiting the inevitable process of urbanization and industrialization. As a result, top-down rural planning in Canada is seen as disjointed, inconsistent, paternalistic and insensitive to local needs.

Top-down policy responses tend to be sectoral in nature and therefore often cross-cut the more holistic problems of deprivation and disadvantage in rural communities. The principal policy sector is that of agriculture, and without exception the case study chapters have stressed the importance of central state agricultural policies in the various nations under review. Whereas other industrial sectors have undergone capital restructuring with only indirect support from the state, agriculture has been the recipient of major state financial subsidies. As a consequence, the restructuring of agriculture has been inextricably intertwined with political power relations within the state. In many cases this has meant that once agricultural policy support systems had been put into place they have been difficult to remove or adjust in response to the structural imperatives of international markets. Policies designed to ensure a desired level of production in particular crops have tended to resist potential reforms which have been prompted by a recognition of overproduction, unequal product competition and other market-oriented circumstances.

To a significant degree, the resistance to policy restructuring in the agricultural sector has been due to the powerful political lobby enjoyed by agricultural interests in most mixed economies. This in turn has led to central government departments with responsibility for agriculture – the US Department of Agriculture, the Ministry of Agriculture, Fisheries and Food in Britain, and so on – assuming a lead rôle in the promotion of well-being in the rural sector. *Rural* welfare thus becomes *agricultural* welfare in many respects.

Although there is evidence in many nations that the benefits of agricultural support are unevenly distributed, not only amongst rural populations but also within the farming fraternity, it has generally not been the plight of the small, low-income farmer which has generated policy responses to the need for structural change in agriculture. Apart from nations such as France where the Common Agricultural Policy of the European Economic Community is so scheduled as to prop up the incomes of certain categories of smaller farmers and where small-scale farmers do wield some political clout, it has been the larger-scale political issues of central state expenditure, balance of payments, and specifically the over-burdening *cost* of agricultural support which has prompted the (usually slow) changes in agricultural policy. For example, John Holmes's account of agricultural policies in Australia demonstrates how government assistance is increasingly being directed towards adjusting production to low world prices and unequal competition from the USA and the European Community.

There are signs, however, that the favourable treatment enjoyed by agricultural capital is beginning to wane in some nations. Born of strategic necessity and of a perceived economic bond between indigenous agricultural production and national stability, the agricultural sector has sucked in significant proportions of central state discretionary expenditures. In Britain, for example, it has been estimated that two-thirds of the gross profits from agriculture are derived from the national and European states. Inevitably, other fractions of capital have aimed to climb onto this bandwagon, either by prising some of these expenditures away from agriculture and into support systems favouring their own production sectors, or by reducing government expenditure overall so that their own taxation burdens are correspondingly lightened. One possible crisis of capital is hypothesized as stemming from just such conflicts between different capital fractions.

The beginnings of a rift in this privileged relationship between government and agricultural interests are apparent in Britain, where the Minister of Agriculture was recently given an unprecedented vote of censure by the National Farmers Union over the impacts of EEC measures to reduce production in some products and, more especially, over proposed changes to agricultural land-use planning. The most radical case of such a rift, however, has been in New Zealand. In 1984 the newly elected Labour government removed farm income supports and subsidies with a resultant decline in land prices and agricultural incomes.

The impact of any diminution of agricultural bias in rural policies will depend on the degree to which central states merely withdraw policy and financial support from rural areas, or whether resources are substituted into a planned adjustment of rural economies in response to agricultural restructuring. In some advanced and more densely populated nations like Britain and the peri-urban parts of most nations discussed in this book, we may be moving towards a phase of contemplating the requirements of a post-modernist agriculture. If so, a key question is whether the strategic need for a supported *rural* population will prove as compelling to governments as the strategic need for an *agricultural* population has done during the age of industrial progress.

A second type of top-down sectoral policy affecting rural areas is regional policy. This generic term has been used to describe many different policy packages, but in its most basic form, regional policy represents the provision by central governments, such as those in Scandinavia described by Peter Sjoholt, of economic incentives to industrial enterprises to move from growth areas to the periphery, alongside efforts to promote and support indigenous economic activities in the marginal areas. In many instances the 'margin' or 'periphery' will overlap with rural areas, but it should be noted that regional policies also (and often principally) apply to declining industrial areas, which happen to be on the economic rather than spatial margins of the time. Indeed, the political impetus for regional policies has often stemmed from a need for high-profile government action in these old industrial areas. Once again then, top-down policies, which are initiated for the support of particular interests, have been relevant to rural areas and legitimated as responses to rural problems.

Another example of regional industrialization policies is the use of major resource-based projects to stimulate economic activity in marginal areas. Examples of such schemes abound, for example, in the account of Australian policies, and in Richard Willis's description of New Zealand's so-called Think Big programme. Here central government has implemented a policy of developing the nation's energy resources so as to achieve greater self-sufficiency in energy. Such projects as the Taranaki gas developments have sponsored a population boom and at least short-term economic spinoffs for a rural area. Again the point is made, however, that this is a case of *national* economic policy affecting rural areas rather than policies which are specifically designed as a response to rural problems.

The evidence presented in this book does illustrate some regional planning schemes which are directed to particular rural areas and which offer a package of policies which may be interpreted as more of a response to particular problems in the areas concerned. In the USA, for example, specific rural areas have been regarded as depressed regions and programmes for the Tennessee Valley and the Appalachian region have been implemented in this context. These programmes have included a strong bias towards the development of growth centres (see below) in the hope that the benefits of growth in key urban settlements would trickle down to the rural areas.

To some extent the specific planning initiatives within rural regional policies are dictated by scale. In the USA, the Tennessee Valley Authority had jurisdiction over seven states, and the Appalachian Regional Planning Commission covered a thirteen-state region. Emphasis has been on resource development and the improvement of physical infrastructure so as to attract new industrial development. As Lassey and his colleagues ruefully suggest, the provision of new infrastructure, particularly that connected with transportation, can facilitate the *exit* of people and jobs as well as the intended entry.

Where regional schemes are implemented over smaller areas, such as the Highlands and Islands of Scotland and Mid Wales in Britain, more specific measures have been included to attract new economic enterprise so as to help restructure ailing rural economies. Even with the provision of site and building infrastructure and the offer of subsidies and loans to potential in-migrant and indigenous enterprises, the development boards in Britain have, like their counterparts in the USA, had to concentrate on key centres as the major recipients of growth, thus leaving themselves open to the criticisms that if trickle-down effects cannot be achieved, then the beneficiaries of regional planning activities tend to be the capital interests (who gain from subsidies) and the citizens of the growth centres (who gain job opportunities) but not the rural residents of the hinterland, who often do not gain from centralized development.

Hugh Clout's account of area-based planning in France portrays a type of smaller-scale regional planning based on the establishment of rural action zones and rural renovation zones to which special development funds were allocated. Later initiatives in the Massif Central were emphasized by the French government to be based on the political imperative of maintaining a population in remote rural areas as well as on economic and environmental arguments.

One key point to comparison regarding regional planning in rural areas is that none of the nations analysed in the case studies has established a *comprehensive* regional planning system covering all rural areas. The selectivity of regional programmes might be viewed as a financially derived necessity, but it also reflects that although the rural problems which underlie regional planning measures appear to be widespread the response to these problems is prompted more by the political need to act (perhaps to deflate political opposition such as from nationalist groups in Scotland and Wales) than by a recognition that particular problems in all needy areas should be responded to by planning and other policies.

One further element of top-down policy making for rural areas concerns the service sector. In some senses rural services are protected in an unannounced and often unexamined manner. In Britain, for example, charges made for postal services, telephone calls (although this is now changing with the privatization of British Telecom), gas, electricity and water services are either standard or of the same level of magnitude for both rural and urban residents. Price standardization in fact represents a subsidy to rural residents whose facilities are more costly to install and maintain.

Several explicit illustrations of service sector policies generated by central governments have arisen in the case study chapters. For example, the analysis of government policy in France by Clout demonstrates a specific initiative in rural service policy from the central state. In 1974, the government announced the requirement of a six-month warning before any rural service in any *département*, and in the public sector services under their control, such as post offices, village junior schools and secondary schools, the critical population thresholds for servicing were all revised downwards. This move contrasts with recent government advice to local authorities in Britain, which makes the closure of small rural facilities increasingly likely on grounds of economic efficiency during a period of public sector expenditure cuts.

In Australia, and other nations including expansive remote rural areas, more innovative service policies have been introduced in order to maintain a population presence in marginal areas. In terms of education, the one-teacher school, peripatetic teachers, correspondence schools and the School of the Air have all been used to ensure that residents of isolated areas are included in state service provision. Comparable schemes, such as the Royal Flying Doctor service and the lower-profile community health projects, have been introduced in the health care sector.

Again, the more general provision of public sector service facilities in rural growth centres has been instrumental in providing employment and generating other economic activities in country towns. There is some evidence that it is this public sector investment in the spread of local government, health and education services which has been at least partially responsible for the population upturn that has been experienced in the more peripheral areas of European nations such as Britain and the Netherlands. As with other top-down policies, however, the intentions behind these policies form part of government's attitude towards welfare delivery. The seemingly pro-rural service policies have to be measured against the losses of services which have taken place particularly in the private sector, and indeed against the propensity to privatize key service delivery agencies. Arguments of equity versus efficiency are paramount at this point, but it could be argued that service provision represents one spoke in the umbrella of state legitimation policies rather than a direct response to the needs of rural people. These concepts further underlie another group of comparable public policies, namely the designation and development of rural growth centres. It is to this middle-ground policy area that we now turn.

(2) MIDDLE-GROUND POLICIES FOR GROWTH CENTRES

One seemingly universal response to the question of rural policy making has been the designation of growth centres to act both as economic servicing nodes and as points of intervening opportunity which serve to interrupt out-migration flows. Such policies fall into a middle ground of state policy making, because although central governments have in some cases advised regional and local state agencies on the merits of growth-centre policies, it has often been these lower-tier governments which have formulated detailed policies for rural settlement patterns.

Perhaps the most marked example of these policies has been implemented in the USSR. Pallot's chapter has described the deliberate decision by the Soviet leadership to institute changes in order to raise the quality of rural life. Accordingly, Soviet planners have explicitly sought to restructure the rural settlement network. Settlements have been classified as *perspectivnie* (viable) or *neperspectivnie* (non-viable), and the latter have been left to die out with their populations being encouraged to move out to more viable locations. The scale of the problem of small settlements is enormous, with over half of rural settlements having less than 100 inhabitants, and the Soviet leadership for both ideological and economic reasons required the development of larger centres such as the *agrotowns* in which living standards could be raised closer to urban equivalents.

The implementation of this strategy rested with the formulation of long-term development plans for every rural district. These plans oversee the distribution of the rural settlement network and the location of services and economic activities within it. It was these plans, drawn up by special district-level planning commissions, which classified settlements as viable and non-viable. Initially 85% of villages were classed as non-viable but in 1968 the number of viable villages was raised by one-half because of the obvious social and economic costs of relocation on the original scale. Pallot suggests that the policy of village classification has fallen into official disfavour with the Gorbachev administration. This policy shift is at least partly due to the costs involved with extensive relocation programmes, and so it might be anticipated that any replacement policy (as yet unannounced) will be constrained by cost criteria which will make it difficult to opt for significant levels of resource dispersal throughout the settlement pattern.

On rather obvious and superficial grounds, these policies in the USSR closely resemble the trends of rural settlement planning in mixed economy nations. In the UK, rural settlements have broadly divided into *key settlements* and others, although the classification of individual places has often been far more complex than just these two categories. The settlement strategies have similarly been worked out in county-level development and structure plans which have sought to direct investment into selected centres for economic and social reasons. In his chapter on the Netherlands, Jan Groenendijk demonstrates how the idea of modernizing the settlement pattern by concentrating new development into fewer large villages became ingrained into policy making during the preparation of regional plans by the provinces, and during the provincial assessment of municipal plans. Other similar strategies are apparent in the other case study nations, although the statutory planning instrumentation underlying settlement rationalization varies considerably according to the land-use planning regulations of the nations concerned.

On the surface, then, policies for settlement planning appear curiously similar. It would appear that the modernization and restructuring of the agricultural economy which has traditionally underwritten all rural societies inevitably brings in its wake an economic necessity for the rationalization of settlement patterns. As the insularity and isolation of rural society *per se* becomes broken down through labour transfers into other economic sectors and with the establishment of mass communications diffusing urban cultures and life-style ambitions, the willingness of rural people to accept the peasant way of life is eroded. Ambitions for better standards of living, especially in terms of service delivery, become a political issue, and governments are faced with the task of providing facilities and services within cost imperatives. The economies of scale, and sheer pragmatic benefits in both financial and administrative spheres, offered by resource-concentration policies appear to have been irresistible in the development of policies for rural people.

A closer examination of these apparently similar policies, however, reveals significant variations in the political economic conditions which shape the development and implementation of growth-centre policies. In the USSR, the recognition of rural viability is underpinned by motives of maintaining an efficient and sufficient workforce in rural areas for the cultivation of a gradually restructuring agricultural economy, and of the eradication of poverty according to communist ideologies. Obviously the manner of implementation and decision making relating to settlement policies in the USSR is fundamentally different to that in non-communist nations. Having said that, it is possible to point to the demolition of a few old mining villages in the county of Durham in the UK (nicknamed Durham's 'D' villages) and suggest, with some validity, that in terms of policy outcome and public reaction to the impact of strictly administered key settlement policies there are some significant similarities in the *potential* of these policies in different political environments. The difference is that this isolated example of the use of bulldozers to restructure the settlement pattern was not replicated throughout the settlement system in Britain. Other chapters, however, have described the 'ghost towns' which have resulted from economic trends of decline which have not been regulated by the intervention of public policies of economic support and resource provision. Whether by dint of the bulldozer, or by conscious political decisions to allow settlements to die, there are interesting comparisons to be made concerning the mechanisms and after-effects of rural settlement rationalization policies across the spectrum of political economies.

The impact of rural settlement policies will also depend on the degree to which conditions in non-viable settlements can be cushioned, either by market forces or by planning intervention. For example, in expansive geographical margins such as those in Canada, Australia and especially the USSR, sheer scale dictates that alternative residential uses of economically non-viable settlements are unlikely. This situation is different from that in the more densely populated nations in Europe (for example, the UK, the Netherlands and parts of France and Scandinavia) where small settlements which are reasonably accessible to urban labour markets have become gentrified with the influx of adventitious middle-class and especially service-class in-migrants. In such locations, rural planning policies are also influenced by the need to conserve the architectural and environmental heritage of the rural settlementscape. Such aspects are politically self-reinforcing, as the middle-class newcomers have exerted strong local political pressures to preserve the environmental integrity and the invest-ment value of their chosen rural residences. Planners' attempts to introduce policies of restricting new residential development in these villages to that required for 'local need' have been relatively unsuccessful because land-use planning regulates the type of house which may be built but *not* the type of residents who live there. 'Local needs' development, therefore, often provides small additions to the housing stock to be gobbled up by the voracity of gentrification processes.

(3) BOTTOM-UP COMMUNITY POLICIES

Brief mention should also be made of the increasing trend within rural policies towards the incorporation of local community action into the for-mal planning process. In the case of Canada, for example, Hodge reports a surge of community self-reliance efforts in small rural communities. The reason behind this phenomenon is clearly stated to be the shortfalls in traditional strategies of development – that is more formal policies from government agencies for the economic viability of small rural com-munities. Hodge characterizes these small communities as waiting for out-side investment that does not materialize so instead the communities set about their own economic development projects, often with the encouragement of provincial planners and governments. Similarly, community organizations are shouldering increased responsibility for the delivery of local social services to rural people.

This trend towards self-help was also highlighted in the chapter on rural Britain. It is not, however, the self-help phenomenon itself which is significant. Rural communities in nations such as New Zealand have been avid self-helpers since time immemorial, and in most of the nations studied in this book, rural self-help has existed as a backcloth to community activity regardless of official policies and plans. The important shift of circumstances from a rural policy viewpoint is the increasing propensity for self-help and voluntary activity to be actively supported and espoused by such official policies. In many nations, official policy statements now applaud the partnership which is being created between statutory planning, development and service delivery on the one hand and community initiatives on the other. Such developments are even shadowed in the USSR, where the development of clubs, sport centres, shops and public buildings has been brought about in small rural settlements as a result of joint ventures between local councils, farm managements and individual rural families.

The partnership of public sector policy and community initiative can be viewed from two contrasting perspectives. First, this trend can be seen as a truly bottom-up process of significant democratic validity. On the grounds that local people will be more sensitive to the particular needs of their own community than will detached bureaucrats and politicians in some far-off political centre, community-based action represents a valid and worthwhile function. When the recognition of need has to be converted into responsive action, there are indeed many areas (such as citizens' advice, neighbourly help and so on) where state intervention is both unwanted and unnecessary. Furthermore, some communities do have the skills and financial resources to fund their own initiatives, thereby ensuring the implementation of local responses. Others, however, do not have these skills and resources, or even the necessary qualitites of initiative and community leadership. A general policy of reliance on self-help will therefore discriminate against these types of communities, and there does seem to be an aggregate negative correlation between communities who are most *able* to help themselves and communities who are most *in need* of help.

A second perspective on community partnership interprets government espousal of self-help as an abrogation of governmental responsibilities. In other words, by making official the increasing expectation that community initiative represents one (sometimes the only) method of responding to the needs of rural people, governments may be seen as indulging in a political cop-out. Services and facilities which have previously been provided by the public sector, or which might reasonably be expected to be financed from the public purse, are now being consistently and detrimentally affected by central and local state crises of expenditure. In the absence of public funds, communities are being expected to substitute their own resources into local service provision. For planners this may be the only resource open to them for bringing about a response to rural problems in a political and economic environment where planning intervention has been reduced to planning by opportunism. In the political decision-making sphere, however, this policy switch represents a rolling back of the welfare state, the consequences of which will be dramatic as we move into the 21st century.

Planning as a state function

The response of nation–states to prevailing economic conditions and crises has certainly had an impact upon the policies and plans for rural people. Different states have reorganized power relationships and political administration in different ways. Britain stands out as a nation where increasing power has accrued to the central state, despite the fact that the local state continues to function as the major agent of service delivery within the public sector. By exerting increasing control over the local state through regulation of expenditure and legality of action, the central state in Britain has set about achieving major economic objectives by centralized direction. The blanket centralization has weakened the discretionary powers of the local state, and thereby weakened the powers of responsive planning and policy making at the local level. Even if more autonomy was enjoyed by local decision-makers, however, it is unlikely that higher levels of planning intervention on behalf of needy rural groups would be contemplated because of the innate conservative nature of local rural politics. A somewhat similar configuration of political power is currently emerging in other European nations such as the Netherlands, where central state restrictions on public expenditure and a conservative local state are important factors in the interpretation of rural policies.

An alternative response to economic and political change has been to decentralize power within the state. In France, for example, Clout reports that the Socialist administration has attempted to break away from administrative centralization by giving new powers to the 22 *regions* and granting additional responsibilities to the 36 512 *communes* geared towards a greater level of management of their own resources. This decentralization has had the effect of accentuating the importance of rural (or, perhaps more accurately, agricultural) issues but depends for its impact on these problems on the degree to which administrative decentralization is accompanied by the provision of sufficient resources for expenditure by localized planning agencies. This dilemma is also illustrated by the example of the USA. Here small town and county governments have found themselves with increased responsibilities but with declining revenues with which to finance service delivery. In particular, the decline in federal and state revenue sharing has had a detrimental impact on rural jurisdictions, which find themselves facing unequal competition from urban administrations for limited grant funds in an increasingly grant-oriented public economy.

In effect, the centralization/decentralization responses by nation–states have had similar outcomes in that the capacity of local agencies to fund planning interventions is being squeezed. This economic context for planning and policy making within the state is being consistently allied with government trends of handing over previously planned functions to the market-oriented direction of the private sector – the *deregulation* issue. In the example of the USA, Lassey *et al.* describe a rush to deregulate industry with the result that private sector service providers (for example, freight transport, air transport and bus transport) have abandoned rural areas in favour of the higher profits generated in urban environments. Deregulation of public transport services and privatization of telecommunications and gas services in Britain have prompted similar worries as have the proposals to privatize public service monopolies in Australia. In effect, the private sector is being increasingly permitted to take over the planning and policy making for many essential services and developments in the rural areas of many nations. In the advanced capitalist states, which form the bulk of case-study evidence in this book, the private sector and its market orientation have always been important factors in the provision of rural facilities. The recent shift towards freedom of action for the private sector and away from public sector regulation of private capital marks an important trend in the pursuance of wider objectives by the state in capitalist nations.

The impacts of this shift have led to differing configurations of private–public sector relations in rural areas. In broad terms, private sector organizations are becoming increasingly significant as decision-makers. Their involvement in the development of facilities in rural areas may take the form of a partnership with government. In France, for example, mixed-economy corporations, combining public and private sector investment, have been established to provide integrated responses to rural problems. It is interesting to note Clout's reservations as to the success of their activities. Elsewhere private sector involvement takes place in the context of unregulated free marketism. Willis demonstrates that in New Zealand free-market policies pursued by the new Labour government have led to an explosion of economic activity in the cities, but the rural areas have suffered from the same policies as deregulation strategies have exerted a detrimental impact on farmers and rural services.

If rural development is to be increasingly demand-led, rather than planned as a response to need, then the account given by Lassey *et al.* of Big

Sky, Montana – the purpose-built new town resort – will be repeated elsewhere. Profitable projects for affluent groups will prosper; interventionist action by government on behalf of non-affluent groups will not.

Part of the relative failure of public sector intervention in rural areas appears to be due to a lack of integration of agencies and policies concerned with rural territory and people. Hodge's account of policies in Canada characterizes rural agencies as being prone to 'duplication, interdepartmental conflict and competition among programs', and Holmes's view of Australian policy suggests that the lack of involvement in comprehensive or integrated planning is a reflection 'in part of lack of any urgent need for such planning, combined with an institutional framework which is ill-suited to such initiatives'. Similar problems are experienced in most of the case study nations.

It does seem fair to conclude, therefore, that policies and plans for rural people are being fashioned by the various interrelations mentioned at the end of the chapter on the United Kingdom:

- central–local relations,
- private–public sector relations, and
- interagency relations.

Evidence of conflict in each of these spheres appears with regularity and significance in the case-study chapters. But what of the primary relations between state and society? Does the evidence presented in this book throw light on the overriding function of the state with regard to the key questions of neutrality or activity on behalf of dominantly powerful classes and interests? In fact few of the authors have given specific analysis to this question, but it is evident from several of the case-study chapters that planning mechanisms are being used for underlying purposes by the state. For example, Sjoholt's description of growth-centre policies in Scandinavia notes that

> the local physical planning authorities were used systematically as instruments in facilitating labour mobility from rural problem areas. Simultaneously, the municipal planning institutions helped in easing this mobility, seeing that new infrastructure was provided to accommodate this labour force in the centres of growth.

This view of planning as an instrument of facilitation is one that is implicit in much of the material presented in this book. But the underlying state function which motivates facilitation is a matter for interpretative analysis rather than crude verification or falsification from case-study evidence.

In the introductory chapter I outlined three brief snapshots of different conceptual perspectives on this complex question of state function:

(1)   market-oriented state intervention on behalf of the individual;
(2)   the state on the sidelines, intervening for the common good;
(3)   state intervention on behalf of capital and class.

This book has shown that any holistic view of the form, function and apparatus of the state tends to overlook the important complexity of interests and bargaining that occur in and around the planning process. There does seem to be important analytical value in understanding the state as an agency for the protection and furtherance of capital accumulation, and therefore of capitalist-class interests. In addition, however, we must take note both of how individual capitalist nation–states have developed different mechanisms and apparatus to perform this overall function, and of how the socialist state of the USSR has opted for a strikingly familiar formula for its activities of settlement reorganization in rural areas. The state has therefore to be viewed as simultaneously pursuing its support of capital interests *and* carrying out autonomous and discretionary activities as the occasion arises. Thus it can be shown that intervention in rural areas has benefited certain classes and interests, and that the espousal of free-market objectives and individual freedoms has contributed to that objective (even if the 'freedoms' of the rural deprived are not of the same quality as those of the rural wealthy). At the same time, government policies to provide basic services and infrastructure in rural areas cannot merely be discounted as explicit political tactics of legitimation, since different states seem to have gone beyond strategic or political control requirements in their servicing policies for small groups of people in marginal areas.

Cliff Hague (1984: 43) has written:

> Thus the crises of capitalism are written in the settlement structure and the built environment, and that structure and environment become the focus for political struggles pitched with varying degrees of consciousness at resisting the offerings and imperatives of capital.

It can be strongly argued that studies of rural planning and policy making have for too long ignored the crises of capitalism which are written into the rural settlement structure and environment. Understanding of rural change in the future will inevitably increasingly entail understanding the imperatives of organized or state capital, rather than seeking out superficial policy responses to perceived rural needs. Yet these common factors reflecting a *service state* which can be seen to underlie the case studies presented in this book must be set alongside the contrasts in apparatus and mechanism employed by a *discretionary state* in its dealings with rural people. This difficult dualism of servility and discretion holds the key to an understanding for policies and plans for rural people in the developed world.

References

Cherry, G. E. (ed.) *Rural planning problems*. London: Leonard Hill.
Hague, C. 1984. *The development of planning thought*. London: Hutchinson.

Index